Mathematical Programming for Power Systems Operation

Mathematical Programming for Power Systems Operation

From Theory to Applications in Python

Alejandro Garcés
Technological University of Pereira
Pereira, Colombia

This edition first published 2022

Published by John Wiley & Sons, Inc., Hoboken, New Jersey.
Published simultaneously in Canada.

For general information on our other products and services or for technical support, please contact our Customer Care Department within the United States at (800) 762-2974, outside the United States at (317) 572-3993 or fax (317) 572-4002.

Wiley also publishes its books in a variety of electronic formats. Some content that appears in print may not be available in electronic formats. For more information about Wiley products, visit our web site at www.wiley.com.

Library of Congress Cataloging-in-Publication Data
A catalogue record for this book is available from the Library of Congress

Paperback ISBN: 9781119747260; ePub ISBN: 9781119747284;
ePDF ISBN: 9781119747277; oBook ISBN: 9781119747291

Cover image: © Redlio Designs/Getty Images
Cover design by Wiley

Contents

Acknowledgment

Throughout the writing of this book, I have received a great deal of support and assistance from many people. I would first like to thank my friends Lucas Paul Perez at Welltec, Adrian Correa at Universidad Javeriana in Bogotá-Colombia, Ricardo Andres Bolaños at XM (the transmission system operator in Colombia), Raymundo Torres at Sintef-Norway, and Juan Carlos Bedoya at the Pacific Northwest National Laboratory (USA), who, in 2020 (during the COVID-19 pandemic), agreed to discuss some practical aspects associated to power system operation problems. The discussions during these video conferences were invaluable to improve the content of the book. I am also very grateful to my students, who are the primary motivation for writing this book. Special thanks to my former Ph.D. students, Danilo Montoya and Walter Julian Gil. Finally, I want to thank the Department of Electric Power Engineering at the Universidad Tecnológica de Pereira in Colombia and the Von Humbolt Foundation in Germany for the financial support required to continue my research about the operation and control of power systems.

Alejandro Garcés

Introduction

Electrification is the most outstanding engineering achievement in the $20th$ century, a well-deserved award if we consider the high complexity of generation, transmission, and distribution systems. An electric power system includes hundreds or even thousands of generation units, transformers, and transmission lines, located throughout an entire country and operated continuously 24 hours per day. Running such a complex system is a great challenge that requires using advanced mathematical techniques.

All industrial systems seek to increase their competitiveness by improving their efficiency. Electric power systems are not the exception. We can improve efficiency by introducing new technologies but also by implementing mathematical optimization models into daily operation. In every mathematical programming model, we require to perform four critical stages depicted in Figure . The first stage is an informed review of reality, identifying opportunities for improvement. This stage may include conversations with experts in order to establish the available data and the variables that are subject to be optimized. The second stage is the formulation of an optimization model as given below:

$$\min \ f(x)$$
$$\text{subject to } x \in \Omega \qquad\qquad (0.1)$$

Where x is the vector of decision variables, f is the objective function and, Ω is the set of feasible solutions. Going from stage one (reality) to stage two (model) is more of an art than a science. One problem may have different models and different degrees of complexity. Practice and experience are required to master this stage, as some models are easier to solve than others. Subsequently, the third stage consists of the implementation of the mathematical model into a software. After that, the fourth stage is the analysis of results in the context of the real problem.

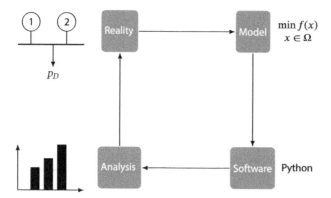

Figure 0.1 Stages of solving an optimization problem.

This book will focus on stages two and three, associated with power system operations models. In particular, we are interested in models with a geometric characteristic called *convexity*, that present several advantages, namely:

- We can guarantee the global optimum and unique solution under well-defined conditions. This aspect is interesting from both theoretical and practical points of view. In general, a global optimum advisable in real operation problems.
- There are efficient algorithms for solving convex problems. In addition, we can guarantee convergence of these algorithms. This is a critical aspect for operation problems where the algorithm requires to be solved in real-time.
- There are commercial and open-source packages for solving convex optimization models. In particular, we are going to use CvxPy, a free Python-embedded modeling language for convex problems.
- Many power system operations problems are already convex; for example, the economic and environmental dispatches, the hydrothermal coordination, and the load estimation problem. Besides, it is possible to find efficient convex approximations to non-convex problems such as the optimal power flow.

In summary, convex problems have both theoretical and practical advantages for power systems operation. This book studies both aspects. The book is oriented to bachelor and graduated students of power systems engineering. Concepts related to power systems analysis such as per-unit representation, the nodal admittance matrix, and the power flow problem are taken for granted. A previous course of linear programming is desirable but not mandatory. We do not pretend to encompass all the theory behind convex optimization; instead, we try to present particular aspects of convex optimization which are useful in power systems operation. The book is divided into two parts: In the first part, the main concepts of convex optimization are presented, including a distinct chapter about conic optimization. After that, selected applications for power systems operation are presented. Most of the solvers for convex optimization allow mixed-integer convex problems. Therefore, we include models that can be solved in this framework too. The student is recommended to do numerical experiments in order to acquire practical intuition of the problems.

All applications are presented in Python, which is a language that is becoming more important in power systems applications. Students are not expected to have previous knowledge in Python, although basic concepts about programming (in any language) are helpful. Our methodology is based on many examples and *toy-models*. We made a great effort in showing the most simple model with a clean code. Of course, these toy-models are an oversimplification of the real problem; however, they allow us to understand the model and its coding. In practice, we may have complex models that combine different aspects such as the economic dispatch, the unit commitment and/or the optimal power flow. A real operation model may require a sophisticated platform that integrates the model with the supervisory control and data acquisition system (SCADA) operating in real-time. The development of such a real industrial model is beyond the objectives of this book.

1

Power systems operation

Learning outcomes
By the end of this chapter, the student will be able to: • Identify problems related to power systems operation. • Link mathematical optimization models to power systems operation problems.

1.1 Mathematical programming for power systems operation

Mathematical optimization is a fundamental tool for the electrical supply chain, from generation through transmission, distribution, and end-use. It may also be used, in different time frames, from a few milliseconds to several years. This book concentrates on optimization problems for power systems operation. These problems are usually continuous and have a time frame from several minutes to one day. Optimization problems with faster dynamics lie in the control and stability analysis, whereas problems with slower dynamics are planning problems.

Mathematical optimization problems associated with power system operation have existed since the beginning of operations research as an independent area, back in the middle of the 20th century. However, modern technologies such as renewable energies and electric vehicles; and current concepts, such as smart-grids, active distribution networks, and microgrids, have created a renewed interest in mathematical optimization applied to power systems. Smart-grids implicate a massive use of technologies such as power

Mathematical Programming for Power Systems Operation: From Theory to Applications in Python. First Edition. Alejandro Garcés.
© 2022 by The Institute of Electrical and Electronics Engineers, Inc. Published 2022 by John Wiley & Sons, Inc.

electronics, communications, and advanced metering. However, the *smart* aspect of these grids comes from mathematical techniques such as mathematical optimization, that manage these technologies, in order to improve the efficiency, reliability, security, and resilience of the system.

Figure 1.1 depicts schematically four common types of mathematical optimization models. These are linear programming (LP), mixed-integer linear programming (MIP), non-linear programming (NLP), and mixed-integer non-linear programming (MINLP). Another classification is to separate them into convex and non-convex problems. The former include LPs and some NLPs; the latter is the rest of the problems. Convex problems are *well-behaved* in the sense that they have theoretical guarantees, such as global optimum and practical algorithms with fast convergence rate. The first part of the book presents these theoretical aspects. However, not all power systems operation problems are convex; therefore, we need to develop convex approximations for those problems, most of them based on conic optimization as presented in Chapter 5.

A power system is quite complex, and therefore, modeling and implementing mathematical optimization problems are equally complex. We need to gain experience in the complex art of modeling and solving mathematical optimization problems for power system applications. Our approach is to create *toy-models* for each problem. These are simplified models that allow us to understand the central issues and do numerical experiments. In the following sections, we briefly describe each of these toy-models, explained in detail in the second part of the book.

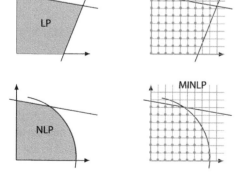

Figure 1.1 Types of optimization models.

1.2 Continuous models

1.2.1 Economic and environmental dispatch

The economic and environmental dispatch of thermal units is one of the most classic problems in power systems operation. It consists of minimizing the operating costs or the total CO_2 emissions, subject to physical constraints such as the power balance and the maximum generation capacity. For the economic dispatch, each generation unit has a cost function f_i, which is usually quadratic and depends on the power generated by each unit. Thus, the objective is to minimize the total cost (or emissions), subject to power balance, as presented below:

$$\min \sum_i f_i(p_i)$$
$$\sum_i p_i = d \tag{1.1}$$
$$p_{min} \leq p_i \leq p_{max}$$

where p_i is the power generated by each thermal unit, and d is the total demand. Environmental dispatch introduces quadratic or exponential functions in the objective function, but the problem's structure is the same. Moreover, power flow constraints can be introduced into the model, although, in that case, it is more precise to name the problem as an optimal power flow (OPF). Chapter 7 presents the economic and environmental dispatches, while Chapter 10 presents the OPF problem.

Another problem closely related to the economic dispatch of thermal units is the unit commitment. This problem considers not only the operating costs but also the start-up and shut-down costs of thermal units. Therefore, the problem becomes binary and dynamic. This problem is studied in Chapter 8.

1.2.2 Hydrothermal dispatch

The economic dispatch problem may include hydroelectric power plants; said plants, generate two fundamental changes into the model. On the one hand, the model becomes dynamic since current operational decisions affect the future operation of the system. On the other hand, the problem becomes stochastic, because the inflows are usually random variables, especially in long-term models. The later aspect is usually solved by an accurate forecasting of the loads and the inflows; hence, it is possible to formulate a deterministic problem, called hydrothermal dispatch or hydrothermal coordination. The basic model has the following structure, namely:

$$\min \sum_t \sum_i f_i(p_{it})$$

$$\sum_i p_{it} + \sum_j p_{jt} = d_t \ \forall t \in \mathcal{T}$$

$$p_{jt} = g(q_{jt}, v_{jt})$$

$$v_{j(t+1)} = v_{jt} + \alpha(a_{jt} - q_{jt} - s_{jt}) \tag{1.2}$$

$$p_{\min} \leq p_{it} \leq p_{\max}$$

$$q_{\min} \leq q_{jt} \leq q_{\max}$$

$$s_{\min} \leq s_{it} \leq s_{\max}$$

$$v_{\min} \leq v_{it} \leq v_{\max}$$

Where i represents thermal units and j enumerates hydroelectric units; t represents the time, thus, p_{it} is the power generated by the thermal unit i at time t. The values of a_{jt}, v_{jt}, p_{jt}, q_{jt}, and s_{jt} are respectively, the inflow, volume, power, outflow, and spillage of the hydroelectric unit j at time t. Figure 1.2, which is self-explanatory, shows these variables.

In this model, g represents the relation between generated power, outflow, and water volume stored in the dam. Although the planning horizon $\mathcal{T}$ may be of short-term (1 day to 1 week), medium-term (1 month), or long-term (1 or more years), we are interested only in the short-term model. As aforementioned, the problem may be stochastic since power demands d_t and inflows a_{jt} are all random variables. However, a determinist model is suitable to understand the problem and its practical implementation. The situation becomes even more problematic when introducing other renewable energies, such as

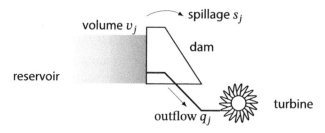

Figure 1.2 Schematic representation of the variables associated to a hydroelectric generation unit.

wind generation and photovoltaic solar generation. Chapter 9 presents the hydrothermal dispatch problem.

1.2.3 Effect of the grid constraints

Both the economic dispatch of thermal units and the hydrothermal dispatch are relatively simple problems. However, the transmission grid introduces additional constraints. Let us consider, for example, a power system with three operation areas, as shown in Figure 1.3. Each line has a maximum power flow capacity that introduces additional constraints in the model (both economic dispatch and hydrothermal dispatch). This constraint limits the flow among areas and modifies the results. These aspects are studied in Chapter 7 from a classic perspective and later, in Chapter 10, under a modern view based on conic optimization.

1.2.4 Optimal power flow

The power flow is one of the most important tools for the analysis of power systems. It allows to determine the state of the power system, knowing the magnitude of voltage and active power in all generating nodes and, active/reactive power demanded in the loads. This results in a non-linear system of equations in complex variables, as presented below:

$$s_k^* = \sum_{m \in \mathcal{N}} y_{km} v_k^* v_m \tag{1.3}$$

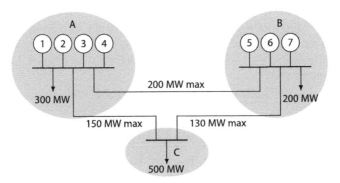

Figure 1.3 Economic dispatch by areas considering network constraints with the transport model.

where $s_k = p_k + q_k$ represents the active and reactive power in node k; $\mathcal{N}$ represents the set of nodes of the grid; y_{km} is the entry km of the Y_{bus} matrix; v_k and v_k are the voltages at nodes k and m, respectively, both represented as complex variables; and s_k^* and v_k^* are the complex conjugate of the respective variables. This representation on the complex number can be splitted into real and imaginary parts. However, as presented in Chapter 10, a complex representation is suitable both for modeling and implementation purposes.

These constraints can be introduced into the economic dispatch, as well as in an optimization model that minimizes total power loss (p_{loss}). In both cases, we named the problem as OPF. The basic model has the following structure:

$$\min_{v,s} P_{loss}(v)$$

$$s_k^* = \sum_{m \in \mathcal{N}} y_{km} v_k^* v_m$$

$$p_{min} \leq p \leq p_{max} \tag{1.4}$$

$$\|s\| \leq s_{max}$$

$$v_{min} \leq \|v_k\| \leq v_{max}$$

$$\text{angle}(v_0) = 0$$

This problem is highly complex due to the non-linear and non-convex nature of the power flow equations. Therefore, it is required to review different approximations that simplify the model. Chapter 10 presents three of these approximations. These are linearization, second-order cone approximation, and semidefinite programming approximation.

Linearization is, perhaps, the most straightforward way to solve the problem. Although the concept of linearization is well-known in real numbers, in this case, we do a linearization on the complex domain. This linearization uses Wirtinger's calculus since Equation (1.3) is non-holomorphic (i.e., it does not have a derivative in the complex numbers). Chapter 10 and Appendix B study this aspect in detail.

We also solve the optimal power flow using second-order cone and semidefinite programming. These approximations demand a basic understanding of conic optimization. Therefore, we present a general background of conic optimization in Chapter 5, and its application to the optimal power flow problem in Chapter 10, including a complete discussion about their advantages and disadvantages.

An optimal power flow may optimize both power systems and power distribution grids. However, the latter case presents some particularities that deserve an independent study. Moreover, both solar and wind energy require power electronic converters connected at power distribution level. These converters are capable of controlling reactive power, and therefore, it is possible to formulate an OPF wherein the decision variables are the power factor of each converter. The problem can be deterministic for real-time operation or stochastic for the day ahead planning. In both cases, the model has, at least, the same complexity as the basic OPF.

1.2.5 Hosting capacity

The concept of hosting capacity refers to the amount of solar photovoltaic generation (or wind) that can be hosted on a power distribution network, at a given time, without adversely impacting safety, power quality, reliability, or other operational features [1]. This analysis can use different performance criteria, including transient and stationary state analysis. The latter alludes to the maximum amount of generated power possible to host without creating over-voltage problems and maintaining operational limits on the distribution lines.

A hosting capacity analysis requires considering the stochastic nature of solar radiation and power demand [2]. Furthermore, it needs to consider the non-linear characteristics related to the power flow equations. Therefore, it is a problem as complex as the optimal power flow (Chapter 11 investigates this problem).

1.2.6 Demand-side management

Demand-side management constitutes a paradigm shift in the context of smart grids, where loads are active components subject to optimization. The role of demand-side management is crucial to decrease CO_2 emissions, reduce the bottleneck in the transmission system, diminish operational cost, and improve efficiency. In order to attain these objectives, it is required to apply mechanisms of electrical load management with static and dynamic techniques. Static techniques involve administrative measures as policies and activities to incentive the end-users to change their energy demand pattern; dynamic techniques include actions to reduce the electricity consumption, such as peak-clipping, valley filling, and load shifting, among others. Information and communication technologies (ICT), as well as the concept of the Internet of the Things (IoT), allow to control the loads and integrate this control in a centralized optimization model.

In general, controllable loads are introduced in a model very similar to the economic dispatch. Its basic structure is presented below:

$$\min \sum_t \sum_i c_{it} p_{it}$$

$$\sum_t (\bar{p}_{it} - p_{it}) \geq d_t \qquad (1.5)$$

$$0 \leq p_{it} \leq \bar{p}_{it}$$

where $\bar{p}_{it}$ is the power required by the load i at time t; p_{it} is the amount of power that is reduced due to the demand-side management model; c_{it} is the cost of disconnecting one unit of power; and d_t is the minimum demand. This is only the basic optimization model, which can be modified, in order to include more type of loads and other aspects of the operation of the system.

Some loads can be moved in time, for example, the washing machine in a residential user. These loads, known as shifting loads, can be optimized by defining the load's optimal starting time. This optimization model is binary but tractable as presented in Chapter 13.

A demand-side management model can also include a model for tertiary control in microgrids or a model for charging electric vehicles. The latter is usually called vehicle-to-grid or V2G. In these cases, the optimization model requires to be executed in real-time by an aggregator as depicted in Figure 1.4.

An aggregator is a crucial component in modern smart distribution networks. This device receives information of the final users – in this case, the electric vehicles – and gives the control actions in order to obtain a *smart* operation. However, the intelligent part of this system is not in the hardware but in the optimization required to solve the problem efficiently and in real-time; therein lies the importance of understanding the optimization model.

A V2G strategy can be unidirectional or bidirectional. In unidirectional V2G, an aggregator controls the electric vehicles' charge similarly as shifting loads.

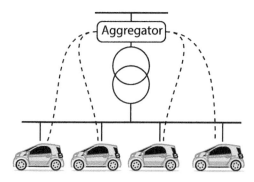

Figure 1.4 Vehicle-to-grid concept with an aggregator that centralizes control actions. Dashed lines represent a communication architecture with the aggregator.

In bidirectional, the electric vehicle can inject power into the grid if required for improving the operation. In any case, the model can become stochastic since the state of charge of the vehicles can be unknown, and the aggregator does not control the arrival/departing time of the vehicles[1]. The aggregator can also incorporate economic dispatch and OPF models to manage other distributed resources such as local batteries, solar panels, and wind turbines. Chapter 11 examines these problems.

1.2.7 Energy storage management

Modern power systems can integrate renewable energy and energy storage devices through a virtual power plant (VPP), an entity that group and centralize the operation of distributed resources to be dispatched by the power system operator. A VPP can encompass an entire region with different renewable sources and energy storage devices. It can also group other microgrids along a distribution feeder.

There are at least two moments where optimization models are required: day-ahead dispatch and real-time operation. Day-ahead dispatch corresponds to the optimization model executed the day before the operation as an economic dispatch model (see Section 1.2.1). This model must include the availability of generation and consider a forecast of the primary resource (inflows, wind, and solar radiance). Moreover, it gives the value power that the VPP operator undertakes on the day of the operation. During the operation, the VPP requires satisfying operative constraints and correcting errors in forecasting the primary resource. Again, a real-time algorithm is necessary for energy storage management.

1.2.8 State estimation and grid identification

The problem of state estimation is classic in power systems. It is also a key component in Supervisory Control And Data Acquisition (SCADA) systems. The problem consists in determining the most probable state of the system from redundant measurements and knowledge of the topology and electrical relations of the grid. When the variables to be measured are active and reactive powers, a non-convex problem is obtained with the same degree of complexity as the load flow. Modern technologies such as the phasor measurement units (PMUs) allow to include direct measures of voltages and angles.

---

1 We incorporate uncertainty in the models using robust optimization. Chapter 6 is dedicated to this aspect.

The problem can be also formulated in power distribution networks and microgrids, both AC and DC. Figure 1.5 shows, for example, a microgrid with a centralized control. Each active element of the network can have both voltage and current measurement. We can use these measurements in order to find the most likely state of the system based on the least squares model as shown below:

$$\min_{I,V} (I - J)^\mathsf{T} M(i - j) + (v - u)^\mathsf{T} N(v - u)$$

$$I = YV \tag{1.6}$$

$$I_{\min} \leq I \leq I_{\max}$$

$$V_{\min} \leq V \leq V_{\max}$$

where J, U are measurements of current and voltage, respectively; I, V are the corresponding estimations and M, N are diagonal matrices that represent the weight of each measurement. The state estimation problem is closely related to the optimal power flow. In fact, some authors call this problem as the inverse power flow problem. The problem is studied in more detail in the second part of the book (Chapter 12).

Another operation problem, closely linked to the state estimation, is the identification of the network. In this case, we have measurements of both voltages and currents at different operating points. Our goal is to estimate the value of the nodal admittance matrix from these measurements. In this case, the optimization model is the following:

$$\min_{Y} f(Y) = \frac{1}{2} \|I - YV\|_F^2 \tag{1.7}$$

The decision variable is the nodal admittance matrix Y, and the objective function is the norm of error between measurements and estimations[2].

The model can include information about the structure of the matrix Y. For example, we already know that the matrix is symmetric and, some of its entries are zero. In that case, the optimization model is the following:

$$\min_{Y} f(Y) = \frac{1}{2} \|I - YV\|_F$$

$$y_{ij} = y_{ji}, \ \forall i, j \tag{1.8}$$

$$y_{ij} = 0, \ \text{if } i \text{ is not connected to } j$$

---

2 It is usual to consider the Frobenius norm as explained in Chapter 12.

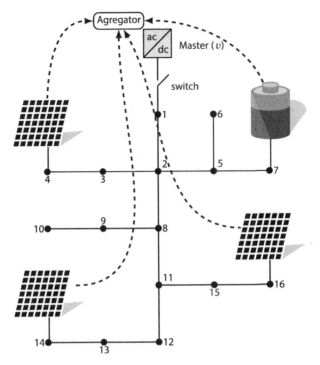

Figure 1.5 Example of a microgrid with a centralized control/measurement in the aggregator.

Both AC and DC grids may handle this type of estimation. In this case, we only presented the DC case because it is easier to develop. The entire model must be implemented in an aggregator structure, as depicted in Figure 1.5.

1.3 Binary problems in power systems operation

Some of the problems previously described are binary. These problems appear in electrical systems, both in the planning and the operation stages. In the planning stage, the problems of transmission expansion planning and distribution planning are typical. Heuristic techniques usually solve these problems together with mixed-integer programming approximations [3]. Although mixed-integer problems are non-convex, they can be solved efficiently using methodologies such as the branch-and-bound algorithm. We present a brief introduction of this method in Chapter 4. Besides, several modules can be called from Python to solve these types of problems. We used Gurobi and Mosek for this task, although there are plenty of options available. Binary operation problems include unit commitment and phase balancing in power distribution networks.

1.3.1 Unit commitment

As aforementioned, the unit commitment problem consists in determining the order of starting and stopping of the thermal units, taking into account the costs of turning-ON and turning-OFF, as well as the starting ramps. The optimization model is similar to the economic dispatch, but in this case, there are binary variables s_k related to the state ON or OFF of each thermal unit. The most simplified model is presented below:

$$\min \ \sum f(p_k) + \sum h(s_k)$$
$$\sum p_k = p_D$$
$$r(s_k) \leq 0 \tag{1.9}$$
$$p_{min} s_k \leq p_k \leq p_{max} s_k$$
$$s_k \in \{0, 1\}$$

where f represents operative costs, h are costs of turning-ON and turning-OFF, r are starting ramps, and p_{min}, p_{max} are the operation limits of each unit. Binary problems are usually difficult; however, it is possible to find either convex approximation or MILP equivalents as presented in Chapter 5.

1.3.2 Optimal placement of distributed generation and capacitors

We can use the same convex approximations of the optimal power flow problem in problems such as the optimal placement of distributed generation in power distribution networks. The problem consists of determining the placement and capacity of distributed generation, subject to physical constraints, such as voltage regulation and transmission lines' capacity. Besides, the problem includes binary constraints related to the placement of the distributed generation. The objective function is usually total power loss, although other objectives, such as reliability and optimal costs, are also suitable.

Another binary problem closely related to the OPF is the optimal placement of capacitors. In this problem, fixed or variable capacitors are placed along the primary feeder to minimize power loss. The capacitors' location and the number of fixed capacitors in each node are binary variables considered in the model. Chapter 11 examines the optimal placement of capacitors and the optimal placement of distributed generation.

1.3.3 Primary feeder reconfiguration and topology identification

The topology of a grid is not constant, especially in power distribution networks. There are several sectionalizing switches along with the primary feeders that permit transferring load among circuits. A feeder reconfiguration is an operating model that determines the optimal topology to minimize power loss. This model is non-linear, non-convex, and binary. In addition, it has a constraint related to the connectivity and radiality of the graph that is tricky to represent in equations. All these aspects are studied in Chapter 11.

Like the state estimation is the inverse of the optimal power problem, there is a problem that can be considered as the inverse of the primary feeder reconfiguration. It is the topology identification in power distribution grids, which takes measurements of current and voltages in different parts of the grid and determines the state of each sectionalizing switch. This problem is studied in Chapter 12.

1.3.4 Phase balancing

Phase balancing is a combinatorial problem that consists of phase swapping of the loads and generators to reduce power loss. Despite being a classic problem, it is still relevant since the unbalance is a common phenomenon in microgrids, especially when single-phase photovoltaic units are included. Because it is a combinatorial problem, phase balancing requires heuristic algorithms with high computational effort. However, it is possible to generate simplified instances of the problem, as presented in Chapter 13.

Each three-phase node has six possible configurations as depicted in Figure 1.6. The problem consists of defining the phase in which each load or generator is connected in order to reduce the grid's total losses. Therefore, the problem is combinatorial since there are 6^n possible configurations, where n is the number of three-phase nodes. The problem is nontrivial even in small

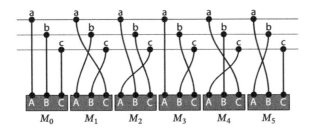

Figure 1.6 Set of possible configurations in a three-phase node.

networks; for example, a microgrid with 10 nodes will have 6×10^7 possible configurations.

Phase-balancing problems appear in many applications such as aircraft electric systems [4], and in power distribution grids with high penetration of electric vehicles [5]. Due to its combinatorial nature, the problem necessitates the use of heuristics [6], and meta-heuristics [7] as well as expert systems [8]. Modern approaches include the uncertainty associated to the load and generator [9].

1.4   Real-time implementation

A receding horizon control can implement most of the optimization algorithms presented in this book. Figure 1.7 depicts the main architecture for this simple but powerful strategy, for real-time implementation of operation models. These optimization models may be the optimal power flow, economic dispatch, energy storage management, or a mixed model that includes multiple models. An unbiased forecast predicts variables such as wind speed, solar radiation, and power demand. Moreover, a state estimator gives accurate measurements of the system variables.

Equation (1.10) represents the optimization model, where x is the vector of decision variables for each time t, and α are the parameters predicted by the forecast module. Of course, this forecast may change since renewable resources and loads may be highly variable in modern power systems. Therefore, the optimization model must be continuously executed and the solution updated.

$$\min f(x, \alpha)$$
$$x \in \Omega \tag{1.10}$$

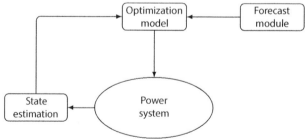

Figure 1.7   A possible architecture for implementing an optimization model for power systems operation.

In many cases, the forecast has an error that introduces uncertainty in the model. Either stochastic or robust optimization is a suitable option to face this uncertainty. Chapter 6 presents the latter option.

1.5   Using Python

Python is a general programming language that is gaining attention in power systems optimization. Although it is neither a mathematical software nor an algebraic modeling language, it has many free tools for solving optimization problems. Moreover, there are many other tools for data manipulation, plotting, and integration with other software. Hence, it is a convenient platform for solving practical problems and integrate different resources[3].

We use a module named CvxPy [10] that allows to solve convex and mixed-integer convex optimization problems; this module, together with NumPy, MatplotLib, and Pandas, forms a robust platform for solving all types of optimization problems in power systems applications. Let us consider, for instance, the following optimization problem:

$$\min c^\mathsf{T} x$$

$$\sum_i x_i = 1 \qquad\qquad (1.11)$$

$$x_i \geq 0$$

where $x \in \mathbb{R}^6$ and $c = (5, 3, 2, 4, 8, 7)^\mathsf{T}$. A model in CvxPy for this problem looks as follows:

```
import numpy as np
import cvxpy as cvx
c = np.array([5,3,2,4,8,7])
x = cvx.Variable(6)
objective = cvx.Minimize(c.T * x)
constraints = [ sum(x) == 1, x >= 0]
Model = cvx.Problem(objective,constraints)
Model.solve()
```

Without much effort, we can identify variables, the objective function, and constraints. This neat code feature is an essential aspect of Python and CvxPy. After the problem is solved, we can make additional analyses using the same platform. This combination of tools is, of course, a great advantage; however, we must avoid any fanaticism for software. There are many programming

---

3 See appendix C for a brief introduction to Python.

languages and many modules for mathematical optimization. What is learned in this book may be translated to any other language. The problem is the same, although its implementation may change from one platform to another.

We made a great effort in making the examples as simple as possible (we call them *toy-models*). This approach allows us to understand each problem individually and do numerical experiments. Real operation models may include different aspects of these toy-models; for instance, they may combine economic dispatch, optimal power flow, and state estimation. These models are highly involved with thousands of variables and constraints. Nevertheless, they can be solved using the same paradigm presented in this book.

Part I

Mathematical programming

2

A brief introduction to mathematical optimization

Learning outcomes

By the end of this chapter, the student will be able to:

- Establish first-order conditions for locally optimal solutions.
- Solve unconstrained optimization problems, using the gradient method, implemented in Python.
- Solve equality-constrained optimization problems, using Newton's method implemented in Python.

2.1 About sets and functions

Sets and functions are familiar concepts in mathematics. A set is a well-defined collection of distinct objects, considered an object in its own right. A function is a map that takes objects from one set (i.e., input or domain) and returns an object in another set (i.e., output or image). An optimization problem consists of finding the *best* object in the output set and its corresponding input, as shown schematically in Figure 2.1. The input set Ω is called the set of feasible solutions, and the best object corresponds to a minimum or a maximum, according to an objective function $f : \Omega \subseteq \mathbb{R}^n \to \mathbb{R}$.

Solving an optimization problem implies not only to find the value of the objective function (e.g., f_{min} or f_{max}) but also the value x, at the input set Ω (e.g., x_{min}, x_{max}). These values are represented as follows:

$$\min_{x \in \Omega} f(x) = f_{min}, \quad \max_{x \in \Omega} f(x) = f_{max} \tag{2.1}$$

$$\operatorname*{argmin}_{x \in \Omega} f(x) = x_{min}, \quad \operatorname*{argmax}_{x \in \Omega} f(x) = x_{max} \tag{2.2}$$

Mathematical Programming for Power Systems Operation: From Theory to Applications in Python. First Edition. Alejandro Garcés.
© 2022 by The Institute of Electrical and Electronics Engineers, Inc. Published 2022 by John Wiley & Sons, Inc.

Figure 2.1 Representation of the sets related to a general optimization problem.

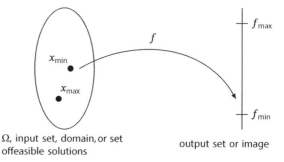

Ω, input set, domain, or set of feasible solutions

output set or image

Notice that f_{min} and f_{max} are numbers whereas x_{min} and x_{max} are vectors. The following example shows the difference between min and argmin (the same applies for max and argmax).

Example 2.1. Let us consider four simple optimization problems and their respective solution, namely[1]:

$$\min(x-5)^2 = 0, \qquad \operatorname{argmin}(x-5)^2 = 5 \qquad (2.3)$$

$$\min_{x \geq 10}(x-5)^2 = 25, \qquad \operatorname*{argmin}_{x \geq 10}(x-5)^2 = 10 \qquad (2.4)$$

$$\min(x-5)^2 + (y-8)^2 = 0, \quad \operatorname{argmin}(x-5)^2 + (y-8)^2 = [5,8] \quad (2.5)$$

$$\min\ \cos(x) = -1, \qquad \operatorname{argmin}\ \cos(x) = \pi \text{ or } 3\pi \text{ or } 5\pi \text{ or } ... (2.6)$$

As aforementioned, the operator min returns a number while the operator argmin can return a vector. Two or more points could produce the same minimum. In that case, the argmin is not unique.

But, what exactly does it mean the best solution? And, what characteristics should have both the sets and the functions involved in the problem? Some mathematical sophistication is required to answer these questions. Finding the best solution in a set implies comparing one element with the rest of the set elements. A comparison is a relation of the form $x \leq y$ or $x \geq y$. However, not all sets allow these types of comparisons; those that enable it are called ordered sets. For instance, the real numbers and the integer numbers are all ordered set. However, complex numbers are non-ordered because such a comparison is not possible (what number is higher: $z_A = 1 + j$ or $z_B = 1 - j$?).

The objective function establishes a criterion of comparison. Therefore, its output must be an ordered set. Nevertheless, the input set may be ordered or non-ordered; it depends on the problem's representation. For example, the optimal power flow in power distribution networks targets an ordered set since

---

[1] At this point, the only tool we have to check these results is plotting the function and locating the optimum.

the active power losses belongs to the real numbers; however, the input may be represented as a vector $(v, \theta) \in \mathbb{R}^{2 \times n}$ or as a set of phasors $(v e^{j\theta} \in \mathbb{C}^n)$. The former is an ordered set, whereas the latter is non-ordered. Other possible representations can be the set of positive definite matrices or positive polynomials, as presented in Chapter 10. In conclusion, the objective function must point to an ordered set, but the input set (i.e., the set of feasible solutions) can be any arbitrary set.

We usually compare values in the output set since our objective is to minimize or maximize the objective function. It is also possible to compare values in Ω when it is an ordered set. However, a comparison between elements of the input set may be different in the output set. A function f is monotone (or monotonic) increasing, if $x \leq y$ implies that $f(x) \leq f(y)$, that is to say, the function preserves the inequality. Similarly, a function is monotone decreasing if $x \leq y$ implies that $f(x) \geq f(y)$, that is to say, the function reverses the identity.

Example 2.2. The function $f(x) = x^2$ is not monotone; for example $-3 \leq 1$ but $f(-3) \nleq f(1)$. Nevertheless, the function is monotone increasing in $\mathbb{R}_+$. In this set, $4 \leq 8$ implies that $f(4) \leq f(8)$ since both 4 and 8 belong to $\mathbb{R}_+$.

An ordered set $\Omega \in \mathbb{R}^n$ admits the following definitions:

- Supreme: the supreme of a set, denoted by $\sup(\Omega)$, is the minimum value greater than all the elements of Ω.
- Infimum: the infimum of a set, denoted by $\inf(\Omega)$, is the maximum value lower than all the elements of Ω.

The supreme and the infimum are closely related to the maximum and the minimum of a set. They are equal in most practical applications. The main difference is that the infimum and the supreme can be outside the set. For example, the supreme of the set $\Omega = \{x : 3 \leq x < 5\}$ is 5 whereas its maximum does not exists. It may seem like a simple difference, but several theoretical analyzes require this differentiation.

Some properties of the supreme and the infimum are presented below:

$$\sup_x f(x) = -\inf_x - f(x) \tag{2.7}$$

$$\sup_x \alpha \cdot f(x) = \alpha \cdot \sup_x f(x) \tag{2.8}$$

$$\sup_x \{f(x) + g(x)\} \leq \sup_x f(x) + \sup_x g(x) \tag{2.9}$$

$$\sup_x \{f(x) + \alpha\} = \alpha + \sup_x f(x) \tag{2.10}$$

Moreover, the last case implies that:

$$\tilde{x} = \operatorname{argmin}\{f(x) + \alpha\} = \operatorname{argmin}\{f(x)\} \tag{2.11}$$

That is to say, the value of x that minimizes the function $f(x) + \alpha$ is the same value that minimizes $f(x)$; for this reason, it is typical to neglect the constant α in practical problems.

Example 2.3. Table 2.1 shows some examples of maximum, minimum, supreme, and infimum.

Table 2.1 Bounds of some ordered sets.

Set	sup	max	inf	min
$\Omega_1 = \{1, 2, 3, 4\}$	4	4	1	1
$\Omega_2 = \{x \in \mathbb{R} : 1 \leq x \leq 2\}$	2	2	1	1
$\Omega_3 = \{x \in \mathbb{R} : 3 < x \leq 8\}$	8	8	3	-
$\Omega_4 = \{x \in \mathbb{R} : 2 \leq x < 9\}$	9	-	2	2
$\Omega_5 = \{x \in \mathbb{R} : 4 < x < 7\}$	7	-	4	-

2.2 Norms

In many practical problems, we may be interested in measuring the objects in a set, either as an objective function or as a way of analyzing solutions. A norm is a geometric concept that allows us to make this measurement. The most common norm is the Euclidean distance given by Equation (2.12)

$$\|x\| = \sqrt{x_1^2 + x_2^2 + \dots x_n^2} \tag{2.12}$$

However, this function is not the only way to measure a distance. In general, we can define a norm as a function $\|\cdot\| : \Omega \to \mathbb{R}$ that fulfills the following conditions:

$$\|x\| > 0 \ \forall x \in \Omega - \left\{\vec{0}\right\} \tag{2.13}$$

$$\|x\| = 0 \to x = \vec{0} \tag{2.14}$$

$$\|\alpha x\| = |\alpha| \, \|x\| \tag{2.15}$$

$$\|x + y\| \leq \|x\| + \|y\| \tag{2.16}$$

The first two conditions indicate that a norm must return a positive value, except when the input is the vector $\vec{0}$. The third condition indicates that it is scalable; for example, the norm must be twice the original vector's norm if we multiply all the vector entries by 2. The last condition, known as the triangle inequality, is a generalization of the triangles' property (therein lies its name).

The sum of any two sides' lengths is greater (or equal) to the remaining side's length. This property is intuitive for the Euclidean norm, but surprisingly it is general for many other functions, such as Equation (2.17):

$$\|x\|_p = \left(\sum_i |x_i|^p\right)^{1/p} \tag{2.17}$$

This function is known as p-norm, where $p \geq 1$. Three of the most common examples of p-norms in $\mathbb{R}^n$ have a well-defined representation, as presented below:

$$\|x\|_1 = \sum_i |x_i| \tag{2.18}$$

$$\|x\|_2 = \sqrt{\sum_i x_i^2} \tag{2.19}$$

$$\|x\|_\infty = \sup_i |x_i| \tag{2.20}$$

The Euclidean distance is equivalent to a 2-norm whereas 1-norm, also known as Manhattan distance, consists in measuring the distance along axes at right angles (see Figure 2.2b), and infinity-norm or uniform norm, takes the maximum distance along axes as shown in Figure 2.2c). In general, $\|x\|_1 \leq \|x\|_2 \leq \|x\|_\infty$. All of these norms are suitable ways to measure vectors in the space.

We can use a norm to define a set given by all the points at a distance less or equal to a given value r, as given in Equation (2.21).

$$\mathcal{B} = \{x \in \mathbb{R}^n : \|x\| \leq r\} \tag{2.21}$$

This set is known as a ball of radius r. Figure 2.3 shows the shape of unit balls (i.e., balls of radius 1), generated by each of the three previously mentioned norms.

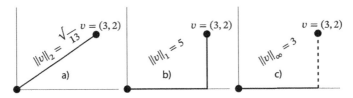

Figure 2.2   Three ways to measure the vector $v = (3, 2)$ in $\mathbb{R}^2$: a) 2-norm or Euclidean norm, b) 1-norm or Manhattan distance, c) infinity-norm or uniform norm.

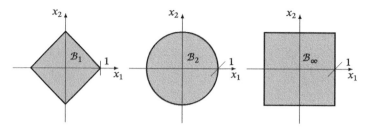

Figure 2.3 Comparison among unit balls defined by norm-2, norm-1, and norm-∞.

Notice that a ball is not necessarily round, at least with this definition. All balls share a common geometric property known as convexity that is studied in Chapter 3.

2.3 Global and local optimum

Let us consider a mathematical optimization problem represented as Equation (2.22).

$$\min_{x \in \Omega} f(x, \beta) \tag{2.22}$$

where $f : \mathbb{R}^n \to \mathbb{R}$ is the objective function, x are decision variables, Ω is the feasible set, and β are constant parameters of the problem.

A point $\tilde{x}$ is a local optimum of the problem, if there exists an open set $\mathcal{N}(\tilde{x})$, named neighborhood, that contains $\tilde{x}$ such that $f(x) \geq f(\tilde{x})$, $\forall x \in \mathcal{N}(\tilde{x})$. If $\mathcal{N} = \Omega$ then, the optimum is global. Figure 2.4 shows the concept for two functions in $\mathbb{R}$ with their respective neighborhoods $\mathcal{N}$.

There are two local minima in the first case, whereas there is a unique global minimum in the second case. This concept is more than a fancy theoretical notion; what good is a local optimum if there are even better solutions in another region of the feasible set? In practice, we require global or close-to-global optimum solutions.

On the other hand, several points may be optimal, as shown in Figure 2.5. In that case, all the points in the interval $x_1 \leq x \leq x_2$ are global optima. Thus, the question is not only if the optimal point is global but also if it is unique. Both globality and uniqueness are geometrical questions with practical implications, especially in competitive markets. Convex optimization allows naturally answering these questions as explained in Chapter 3

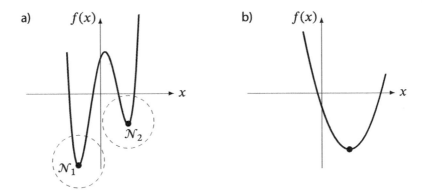

Figure 2.4 Example of local and global optima: a) function with two local minima and their respective neighborhoods, b) function with a unique global minimum (the neighborhood is the entire domain of the function).

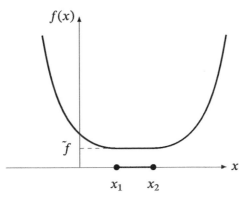

Figure 2.5 Example of a function with several optimal points.

2.4 Maximum and minimum values of continuous functions

It is well-known, from basic mathematics, that the optimum of a continuous differentiable function is attached when its derivative is zero. This fact can be formalized in view of the concepts presented in previous sections. Consider a function $f : \mathbb{R} \to \mathbb{R}$ with a local minimum in $\tilde{x}$. A neighborhood is defined as $\mathcal{N} = \{x \in \mathbb{R} : x = \tilde{x} \pm t, |t| < t_0\}$, with the following condition:

$$f(\tilde{x} \pm t) \geq f(\tilde{x}) \tag{2.23}$$

where t can be positive or negative. If $t > 0$, then Equation (2.23) can be divided by t without modifying the direction of the inequality, to then take the limit when $t \to 0+$ as presented below:

$$\lim_{t \to 0^+} \frac{f(\tilde{x} + t) - f(\tilde{x})}{t} \geq 0 \qquad (2.24)$$

The same calculation can be made if $t < 0$, just in that case, the direction of the inequality changes as follows:

$$\lim_{t \to 0^-} \frac{f(\tilde{x} + t) - f(\tilde{x})}{t} \leq 0 \qquad (2.25)$$

Notice that this limit is the definition of derivative; hence, $f'(\tilde{x}) \geq 0$ and $f'(\tilde{x}) \leq 0$. These two conditions hold simultaneously when $f'(\tilde{x}) = 0$. Consequently, the optimum of a differentiable function is the point where the derivative vanishes. This condition is local in the neighborhood $\mathcal{N}$.

This idea can be easily extended to multivariable functions as follows: consider a function $f : \mathbb{R}^n \to \mathbb{R}$ type $\mathcal{C}^1$ (continuous and differentiable) and a neighborhood given by $\mathcal{N} = \{x \in \mathbb{R}^n : x = \tilde{x} + \Delta x\}$ with $\Delta x \in \mathcal{B}_r$. Now, define a function $g(t) = f(\tilde{x} + t\Delta x)$. If $\tilde{x}$ is a local minimum of f, then

$$f(\tilde{x} + t\Delta x) \geq f(\tilde{x}) \qquad (2.26)$$

In terms of the new function g, Equation (2.26) leads to the following condition:

$$g(t) \geq g(0) \qquad (2.27)$$

This condition implies that 0 is a local optimum of g; moreover,

$$g'(0) = \lim_{t \to 0^+} \frac{f(\tilde{x} + t\Delta x) - f(\tilde{x})}{t} = (\nabla f(\tilde{x}))^{\mathsf{T}} \Delta x \qquad (2.28)$$

Notice that g is a function of one variable, then optimal condition $g'(0) = 0$ is met, regardless the direction of Δx. Therefore, the optimum of a multivariate function is given when the gradient is zero ($\nabla f(\tilde{x}) = 0$). This condition permits to find local optimal points, as presented in the next section. Two questions are still open: in what conditions are the optimum global? And, when is the solution unique? We will answer these relevant questions in the next chapter. For now, let us see how to find the optimum using the gradient.

2.5 The gradient method

The gradient method is, perhaps, the most simple and well-known algorithm for solving optimization problems. Cauchy invented the basic method in the 19th century, but the computed advent leads to different applications that encompass power systems operation and machine learning. Let us consider the following unconstrained optimization problem:

$$\min \ f(x) \tag{2.29}$$

where the objective function $f : \mathbb{R}^n \to \mathbb{R}$ is differentiable. The gradient $\nabla f(x)$ represents the direction of greatest increase of f. Thus, minimizing f implies to move in the direction opposite to the gradient. Therefore, we use the following iteration:

$$x \leftarrow x - t\nabla f(x) \tag{2.30}$$

The gradient method consists in applying this iteration until the gradient is small enough, i.e., until $\|\nabla f(x)\| \leq \epsilon$. It is easier to understand the algorithm by considering concrete problems and their implementation in Python, as given in the next examples.

Example 2.4. Consider the following optimization problem:

$$\min \ f(x, y) = 10x^2 + 15y^2 + \exp(x + y) \tag{2.31}$$

The gradient of this function is presented below:

$$\nabla f(x, y) = \begin{pmatrix} 20x + \exp(x + y) \\ 30y + \exp(x + y) \end{pmatrix} \tag{2.32}$$

We require to find a value (x, y) such that this gradient is zero. Therefore, we use the gradient method. The algorithm starts from an initial point (for example $x = 10, y = 10$) and calculate new points as follows:

$$x \leftarrow x - t\frac{\partial f}{\partial x} \tag{2.33}$$

$$y \leftarrow y - t\frac{\partial f}{\partial y} \tag{2.34}$$

This step can be implemented in a script in Python, as presented below:

```python
import numpy as np
x = 10
y = 10
t = 0.03
for k in range(50):
    dx = 20*x + np.exp(x+y)
    dy = 30*y + np.exp(x+y)
    x += -t*dx
    y += -t*dy
    print('grad:',np.abs([dx,dy]))
print('argmin:',x,y)
```

In the first line, we import the module NumPy with the alias np. This module contains mathematical functions such as sin, cos, exp, ln among others. The gradient introduces two components dx and dy, which are evaluated in each iteration and added to the previous point (x,y). We repeat the process 50 times and print the value of the gradient each iteration. Notice that all the indented statements belong to the for-statement, and hence the gradient is printed in each iteration. In contrast, the argmin is printed only at the end of the process.

Example 2.5. Python allows calculating the gradient automatically using the module AutoGrad. It is quite intuitive to use. Consider the following script, which solves the same problem presented in the previous example:

```
import autograd.numpy as np
from autograd import grad     # gradient calculation

def f(x):
    z = 10.0*x[0]**2 + 15*x[1]**2 + np.exp(x[0]+x[1])
    return z

g = grad(f)   # create a funtion g that returns the gradient
x = np.array([10.0,10.0])
t = 0.03
for k in range(50):
    dx = g(x)
    x = x -t*dx
print('argmin:',x)
```

In this case, we defined a function f and its gradient g where (x, y) was replaced by a vector (x_0, x_1). The module NumPy was loaded using autograd.numpy to obtain a gradient function automatically. The code executes the same 50 iterations, obtaining the same result. The reader should execute and compare the two codes in terms of time calculation and results.

Example 2.6. Consider a small photovoltaic system formed by three solar panels A, B, and C, placed as depicted in Figure 2.6. Each solar system has a power electronic converter that requires to be connected to a common point E before transmitted to the final user in D. The converters and the user's location are fixed, but the common point E can be moved at will. The coordinates of the solar panels and the final user are $A = (0, 40)$, $B = (20, 70)$, $C = (30, 0)$, and $D = (100, 50)$, respectively.

The cost of the cables is different since each cable carries different current. Our objective is to find the best position of E in order to minimize the total cost

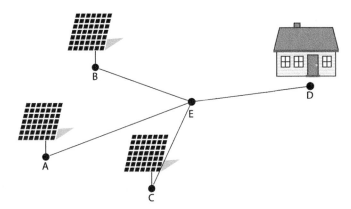

Figure 2.6 A small photovoltaic system with three solar panels.

of the cable. Therefore, the following unconstrained optimization problem is formulated:

$$\min\ f = \text{cost}_{\overline{AE}}\left\|\overline{AE}\right\| + \text{cost}_{\overline{BE}}\left\|\overline{BE}\right\| + \text{cost}_{\overline{CE}}\left\|\overline{CE}\right\|\text{cost}_{\overline{DE}}\left\|\overline{DE}\right\| \quad (2.35)$$

where $\text{cost}_{\overline{ij}}$ is the unitary cost of the cable that connects the point i and j, and $\left\|\overline{ij}\right\|$ is the corresponding length.

The costs of the cables are $\text{cost}_{\overline{AE}} = 12$, $\text{cost}_{\overline{BE}} = 13$, $\text{cost}_{\overline{CE}} = 11$, and $\text{cost}_{\overline{DE}} = 18$. The distance between any two points $U = (u_0, u_1)$ and $V = (v_0, v_1)$ is given by the following expression:

$$\text{dist} = \sqrt{(u_0 - v_0)^2 + (u_1 - v_1)^2} \quad (2.36)$$

This equation is required several times; thus, it is useful to define a function, as presented below:

```python
import numpy as np
A = (0,40)
B = (20,70)
C = (30,0)
D = (100,50)
def dist(U,V):
    return np.sqrt((U[0]-V[0])**2 + (U[1]-V[1])**2)


P = [31,45]
f = 12*dist(P,A) + 13*dist(P,B) + 11*dist(P,C) + 18*dist(P,D)
print(f)
```

The function is evaluated in a point $P = (10, 10)$ to see its usage[2]. The value of the objective function is easily calculated as function of dist(U, V). Likewise, the gradient of f is defined as function of the gradient of dist(U, V) with V fixed, as presented below:

$$\nabla \text{dist}(U, V) = \frac{1}{\text{dist}(U, V)} \begin{pmatrix} u_0 - v_0 \\ u_1 - v_1 \end{pmatrix} \tag{2.37}$$

then,

$$\nabla f(x) = 12 \nabla \text{dist}(x, A) + 13 \nabla \text{dist}(x, B) + 11 \nabla \text{dist}(x, C) + $$
$$18 \nabla \text{dist}(x, D) \tag{2.38}$$

These functions are easily defined in Python as follows:

```python
def g_d(U,V):
    "gradient of the distance"
    return [U[0]-V[0],U[1]-V[1]]/dist(U,V)

def grad_f(E):
    "gradient of the objective function"
    return 12*g_d(E,A)+13*g_d(E,B)+11*g_d(E,C)+18*g_d(E,D)
```

Now the gradient method consists in applying the iteration given by Equation (2.30), as presented below:

```python
t = 0.5
E = np.array([10,10])
for iter in range(50):
    E = E -t*grad_f(E)
f = 12*dist(E,A) + 13*dist(E,B) + 11*dist(E,C) + 18*dist(E,D)
print("Position:",E)
print("Gradient",grad_f(E))
print("Cost",f)
```

In this case, $t = 0.5$ and a initial point $E = (10, 10)$ with 50 iterations were enough to find the solution. The reader is invited to try with other values and analyze the effect on the algorithm's convergence.

The step t is very important for the convergence of the gradient method. It can be constant or variable, according to a well-defined update rule. There are many variants of this algorithm, most of them with sophisticated ways to calculate this step[3]. A plot of $\|\nabla f\|$ versus the number of iterations may be useful for

---

2 Notice P is defined in a line outside the function definition. Recall that x^2 is represented as x**2 in Python (see Appendix C)

3 A complete discussion about the calculation of t is beyond this book's objectives. Interested readers can consult the work of Nesterov and Nemirovskii, in [11] and [12].

determining the optimal value of t and showing the convergence rate of the algorithm, as presented in the next example. We expect a linear convergence for the gradient method, although the algorithm can lead to oscillations and even divergence if the parameter t is not selected carefully. Fortunately, there are modules in Python that make this work automatically.

Example 2.7. The convergence of the algorithm can be visualized by using the module MatplotLib as follows:

```python
import matplotlib.pyplot as plt
t = 0.5
conv = []
E = [10,10]
for iter in range(50):
    E += - t*grad_f(E)
    conv += [np.linalg.norm(grad_f(E))]
plt.semilogy(conv)
plt.grid()
plt.xlabel("Iteration")
plt.ylabel("|Gradient|")
plt.show()
```

The result of the script is shown in Figure 2.7. As expected, the convergence rate is linear; that is to say, the convergence plot describes almost a line in a semi-logarithmic scale. The value of ϵ can be used as convergence criteria (a gradient $\|\nabla f\| \leq 10^{-4}$ can be considered as the local optimum for this problem).

Notice that addition was simplified by the statement +=. In general, an statement such as x=x+1 is equivalent to x+=1. More details about this aspect are presented in Appendix C.

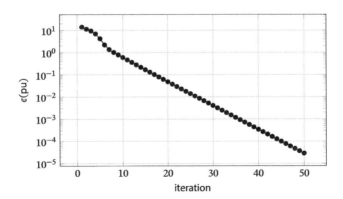

Figure 2.7 Convergence of the gradient method.

2.6 Lagrange multipliers

Reality imposes physical constraints into the problems and these constraints must be considered into the model. For example, an optimization model may include equality constraints, as presented below:

$$\min f(x)$$

$$g(x) = a \tag{2.39}$$

For solving this problem, a function called lagrangian is defined as follows:

$$\mathcal{L}(x, \lambda) = f(x) + \lambda(a - g(x)) \tag{2.40}$$

This new function depends on the original decision variables x and a new variable λ, known as Lagrange multiplier or dual variable. By means of the lagrangian function, a constrained optimization problem was transformed into an unconstrained optimization problem that can be solved numerically, namely:

$$\frac{\partial \mathcal{L}}{\partial x} = \frac{\partial f}{\partial x} - \lambda \frac{\partial g}{\partial x} = 0 \tag{2.41}$$

$$\frac{\partial \mathcal{L}}{\partial \lambda} = a - g(x) = 0 \tag{2.42}$$

by a small abuse of notation, $\partial f / \partial x$ is used instead of ∇f, which is the formal representation for the n-dimentional case, (the same for $\mathcal{L}$ and g). Notice that the optimal conditions of $\mathcal{L}$ imply optimal solution in f but also feasibility in terms of the constraint.

The first condition implies that the gradient of the objective function must be parallel to the gradient of the constraint and, the Lagrange multiplier is the proportionality constant. Besides this geometrical interpretation, Lagrange multipliers have another interesting interpretation. Suppose a local optimum $\tilde{x}$ is found for a constrained optimization problem, and we want to know the sensibility of this optimum with respect to a. The following derivative can be calculated, relating the change of the objective function with respect to a change in the constrain:

$$\frac{\partial \mathcal{L}}{\partial a} = \left(\frac{\partial f}{\partial x}\right)\left(\frac{\partial x}{\partial a}\right) + \left(\frac{\partial \lambda}{\partial a}\right)(a - g(x)) + \lambda\left(1 - \frac{\partial g}{\partial x}\right)\left(\frac{\partial x}{\partial a}\right) \tag{2.43}$$

$$= \left(\frac{\partial f}{\partial x} - \lambda \frac{\partial g}{\partial x}\right)\left(\frac{\partial x}{\partial a}\right) + (a - g(x))\left(\frac{\partial \lambda}{\partial a}\right) + \lambda \tag{2.44}$$

$$= \lambda \tag{2.45}$$

The first two terms in the right-hand side of the equation vanishes, in view of the optimal conditions of $\tilde{x}$; thus, the following expresion is obtained:

$$\lambda = \frac{\partial \mathcal{L}}{\partial a} \tag{2.46}$$

this means that λ is the variation of the lagrangian (and hence the objective function), for a small variation on the parameter a (see Chapter 3 for more details about dual variables).

Example 2.8. Consider the following optimization problem

$$\min x^2 + y^2$$

$$x + y = 1 \tag{2.47}$$

If the problem were unconstrained, the solution would be $x = 0, y = 0$; however, the solution must fulfill the constraint $x + y = 1$. Therefore, a new function is defined as follows:

$$\mathcal{L}(x, y, \lambda) = x^2 + y^2 + \lambda(1 - x - y) \tag{2.48}$$

with the following optimal conditions

$$2x - \lambda = 0 \tag{2.49}$$

$$2y - \lambda = 0 \tag{2.50}$$

$$1 - x - y = 0 \tag{2.51}$$

This is a linear system of equation with solution $x = 1/2, y = 1/2$, and $\lambda = 1$ that constitutes the optimum of the problem.

2.7 The Newton's method

The optimal conditions for both constrained and unconstrained problems constitute a set of algebraic equations $S(x)$ for the first case and $S(x, \lambda)$ for the second case. This set can be solved using Newton's method[4].

Consider a set of algebraic equations $S(x) = 0$ where $S : \mathbb{R}^n \to \mathbb{R}^n$ is continuous and differentiable. Thus, the following approximation is made:

$$S(x) = S(x_0) + \mathcal{J}(x_0)\Delta x \tag{2.52}$$

where $\mathcal{J}$ is the Jacobian matrix of S and $\Delta x = x - x_0$. This constitutes the first-order approximation of S around a point x_0; finding the zero of

---

4 Again, a classic method that became more important with the advent of the computer.

S means to approximate successively the solution by using the following iteration:

$$\Delta x \leftarrow [\mathcal{J}(x)]^{-1} S(x) \tag{2.53}$$

$$x \leftarrow x - \Delta x \tag{2.54}$$

This iteration is the primary Newton's method. Compared to the gradient method, this method is faster since it includes information from the second derivative[5]. In addition, Newton's method does not require defining a step t as in the gradient method. However, each iteration of Newton's method is computationally expensive since it implies the formulation of a jacobian and solves a linear system in each iteration.

Example 2.9. Consider the following optimization problem:

$$\min 10x^2 + 15y^2 + \exp(x + y)$$

$$x + y = 5 \tag{2.55}$$

Its corresponding lagrangian is presented below:

$$\mathcal{L}(x, y, \lambda) = 10x^2 + 15y^2 + \exp(x + y) + \lambda(5 - x - y) \tag{2.56}$$

and the optimal conditions forms the following set of algebraic equations:

$$S(x, y, \lambda) = \left\{ \begin{array}{l} 20x + \exp(x + y) - \lambda = 0 \\ 30y + \exp(x + y) - \lambda = 0 \\ 5 - x - y = 0 \end{array} \right\} \tag{2.57}$$

The corresponding Jacobian is the following matrix:

$$\mathcal{J} = \left(\begin{array}{ccc} 20 + \exp(x + y) & \exp(x + y) & -1 \\ \exp(x + y) & 30 + \exp(x + y) & -1 \\ -1 & -1 & 0 \end{array} \right) \tag{2.58}$$

It is possible to formulate Newton's method using the information described above. The algorithm implemented in Python is presented below:

```python
import numpy as np
def Fobj(x,y):
    "Objective funcion"
    return 10*x**2 + 15*y**2 + np.exp(x+y)

def Grad(x,y,z):
    "Gradient of Lagrangian"
    dx = 20*x + np.exp(x+y) + z
```

---

5 The jacobian matrix of S is equivalent to the hessian matrix of f.

```
    dy = 30*y + np.exp(x+y) + z
    dz = x + y - 5
    return np.array([dx,dy,dz])

def Jac(x,y,z):
    "Jacobian of Grad"
    p = np.exp(x+y)
    return np.array([[20+p,p,1],[p,30+p,1],[1,1,0]])

(x,y,z) = (10,10,1) # initial condition
G = Grad(x,y,z)
while np.linalg.norm(G) >= 1E-8:
    J = Jac(x,y,z)
    step = -np.linalg.solve(J,G)
    (x,y,z) = (x,y,z) + step
    G = Grad(x,y,z)
    print('Gradient: ',np.linalg.norm(G))

print('Optimum point: ', np.round([x,y,z],2))
print('Objective function: ', Fobj(x,y))
```

In this case, we used a tolerance of 10^{-3}. The algorithm achieves convergence in few iterations, as the reader can check by running the code.

2.8 Further readings

This chapter presented basic optimization methods for constrained and unconstrained problems. Conditions for convergence of these algorithms were not presented here. However, they are incorporated into modules and solvers called for a modeling/programming language such as Python. Our approach is to use these solvers and concentrate on studying the characteristics of the models. Readers interested in details of the algorithms are invited to review [12] and [11], for a formal analysis of convergence; other variants of the algorithms can be found in [13]. Moreover, a complete review of Lagrange multipliers can be studied in [14].

This chapter is also an excuse for presenting Python's features as programming and modeling language. The reader can review Appendix C for more details about each of the commands used in this section.

2.9 Exercises

1. Find the highest and lowest point, of the set given by the intersection of the cylinder $x^2 + y^2 \leq 1$ with the plane $x + y + z = 1$, as shown in Figure 2.8.

Figure 2.8 Intersection of an affine space with a cylinder.

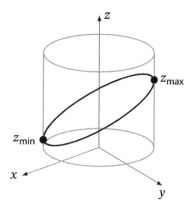

2. What is the new value of z_{max} and z_{min}, if the cylinder increases its radius in a small value, that is, if the radius changes from ($r = 1$) to ($r = 1 + \Delta r$) (Consider the interpretation of the Lagrange multipliers).

3. The following algebraic equation gives the mechanical power in a wind turbine:

$$P = \frac{1}{2}\rho A C_p(\lambda)v^3 \tag{2.59}$$

where P is the power extracted from the wind; ρ is the air density; C_p is the performance coefficient or power coefficient; λ is the tip speed ratio; v is the wind velocity, and A is the area covered by the rotor (see [15] for details). Determine the value of λ that produce maximum efficiency if the performance coefficient is given by Equation (2.60):

$$C_p = 0.73\left(\frac{151}{\lambda} - 13.2\right)\exp\left(\frac{-18.4}{\lambda}\right) \tag{2.60}$$

Use the gradient method, starting from $\lambda = 10$ and a step of $t = 0.1$. Hint: use the module SymPy to obtain the expression of the gradient.

4. Solve the following optimization problem using the gradient method:

$$\min \ f(x_0, x_1) = (x_0 - 10)^2 + (x_2 - 8)^2 \tag{2.61}$$

Depart from the point $(0, 0)$ and use a fixed step $t = 0.8$. Repeat the problem with a fixed step $t = 1.1$. Show a plot of convergence.

5. Solve the following optimization problem using the gradient method.

$$\min \ f(x) = \frac{1}{2}(x - 1_n)^\mathsf{T} H(x - 1_n) + b^\mathsf{T} x \tag{2.62}$$

where 1_n is a column vector of size n, with all entries equal to 1; b is a column vector such that $b_k = kn^2$; and H is a symmetric matrix of size $n \times n$ constructed in the following way: $h_{km} = (m + k)/2$ if $k \neq m$ and

$h_{km} = n^2 + n$ if $k = m$. Show the convergence of the method for different steps t and starting from an initial point $x = 0$. Use $n = 10$, $n = 100$, and $n = 1000$. All index k or m starts in zero.

6. Show that Euclidean, Manhattan, and uniform norms fulfill the four conditions to be considered a norm.

7. Consider a modified version of Example 2.6, where the position of the common point E must be such that $x_E = y_E$. Solve this optimization problem using Newton's method.

8. Solve the problem of Item 4 with the following constraint (use Newton's method):

$$x_0 + 3x_1 = 5 \tag{2.63}$$

9. Solve problem of Item 5 including the following constraint (use Newton's method):

$$1_n^T x = 1 \tag{2.64}$$

10. Newton's method can be used to solve unconstrained optimization problems. Solve the following problem using Newton's method and compare the convergence rate and the solution with the gradient method.

$$\min (x + 3y + 1)^2 + \frac{1}{4}(x - 2y)^4 \tag{2.65}$$

3

Convex optimization

Learning outcomes
By the end of this chapter, the student will be able to: • Identify convex functions, convex sets, and convex optimization problems. • Recognize when a problem has global optimum and unique solution. • Formulate and solve a dual problem.

3.1 Convex sets

The set of feasible solutions of an optimization problem, henceforth feasible set, may have different shapes and properties. It can be open or close, discrete or continuous, linear or non-linear. Each shape determines the type of optimization problem. However, there are sets remarkably well-behaved for optimization problems. These are the convex sets.

A set is convex if we can choose any pair of points within the set so that the line that joins these points also belongs to the set. Figure 3.1 shows an example of a convex and a non-convex set. In the first case, all the points in the line $\overline{x_1 x_2}$ belong to the set. In the second case, there are some points outside the set.

It is easy to identify convex sets in $\mathbb{R}^2$ and even in $\mathbb{R}^3$. We only require to draw the set, and the property becomes evident. However, power systems operation problems are usually in $\mathbb{R}^n$. Therefore, we require a systematic way to identify convex sets.

Mathematical Programming for Power Systems Operation: From Theory to Applications in Python. First Edition. Alejandro Garcés.
© 2022 by The Institute of Electrical and Electronics Engineers, Inc. Published 2022 by John Wiley & Sons, Inc.

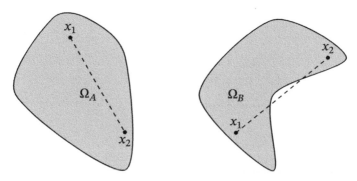

Figure 3.1 Example of a convex set Ω_A and a non-convex set Ω_B.

Example 3.1. Equations and inequalities usually define sets. Let us consider, for instance, the following three sets, related to the 2-norm:

$$\Omega_A = \{(x, y) : x^2 + y^2 \le 1\} \tag{3.1}$$

$$\Omega_B = \{(x, y) : x^2 + y^2 = 1\} \tag{3.2}$$

$$\Omega_C = \{(x, y) : x^2 + y^2 \ge 1\} \tag{3.3}$$

The set Ω_A is convex since any pair of points m, n generates a line whose points belong to the set. The set Ω_B is not convex since it is defined only in the ball's boundary, and hence, any line segment will have points outside the set. The set Ω_C is also non-convex since several points leave the set if we draw, for instance, a line that passes the point $(0, 0)$. Figure 3.2 shows a graphical representation of each of these situations. Notice that we can be misled if we carelessly see the equation.

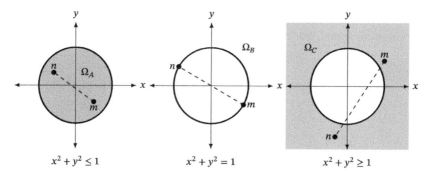

Figure 3.2 Example of three different sets with a similar definition. Only the first set is convex.

We might identify a convex set by a graph in $\mathbb{R}^2$ or $\mathbb{R}^3$, as in the previous example. However, most of the power systems problems are in $\mathbb{R}^n$, where we may not have a simple graphic representation. Therefore, we require a more formal definition, namely: A set $\Omega \subseteq \mathbb{R}^n$ is convex if for any pair of points $x, y \in \Omega$ we have that,

$$(1 - \lambda)x + \lambda y \in \Omega \tag{3.4}$$

for any $\lambda \in \mathbb{R}$, $0 \leq \lambda \leq 1$.

This definition allows determining if a set is convex in a general and systematic way. We only require to select two general points x and y that belong to the set, and demonstrate that any point z in the segment $\overline{xy}$ also belongs to the set too. Let us perform this calculation algebraically, as presented in the following example.

Example 3.2. Let us consider a set defined as $\Omega = \{x \in \mathbb{R}^n : Ax = b\}$. This set is called an affine set. Then, a point $x \in \Omega$ or $y \in \Omega$ must be such that $Ax = b$ and $Ay = b$. Let us consider now a point z between x and y, that is $z = (1 - \lambda)x + \lambda y$, with $0 \leq \lambda \leq 1$. Then we have the following:

$$Ax = b \tag{3.5}$$

$$Ay = b \tag{3.6}$$

$$(1 - \lambda)Ax + \lambda Ay = (1 - \lambda)b + \lambda b \tag{3.7}$$

$$Az = b \tag{3.8}$$

The last equation implicates that $z \in \Omega$ and hence, Ω is a convex set. Notice that we do not select a particular point x or y. Our demonstration was general for any pair of points.

An important property of convex sets is that the intersection of two convex sets generates another convex set. However, the union of two convex is not necessarily convex. Figure 3.3 shows an example of these set-operations.

Figure 3.3 Union and intersection of two convex sets (a triangle and a ball). The intersection is convex but the union may be non-convex.

$A \cap B$ (convex)

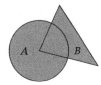

$A \cup B$ (non-convex)

Fortunately, an optimization problem is given by the intersection of the sets generated by their constraints. Consider, for instance, the following optimization problem:

$$\min f(x)$$
$$g(x) \leq 0 \tag{3.9}$$
$$h(x) \leq 0$$

This problem is equivalent to Equation (3.10),

$$\min f(x) \text{ with } x \in \Omega \tag{3.10}$$

where

$$\Omega = \{\Omega_g \cap \Omega_h\} \tag{3.11}$$
$$\Omega_g = \{x \in \mathbb{R}^n : g(x) \leq 0\} \tag{3.12}$$
$$\Omega_h = \{x \in \mathbb{R}^n : h(x) \leq 0\} \tag{3.13}$$

Therefore, it is enough to check that each constrain defines a convex set. We show several classic examples of convex sets below.

Example 3.3. An affine set, is a subset of $\mathbb{R}^n$ given by $\Omega = \{x \in \mathbb{R}^n : Ax = b\}$. This set is convex as demonstrated in Example 3.2. We prefer the term affine instead of linear, since Ω is not a linear space, unless $b = 0$ (a linear space must contain the zero vector [16]).

Example 3.4. A polytope is a set defined as follows:

$$\mathcal{P} = \{x \in \mathbb{R}^n : Ax \leq b\} \tag{3.14}$$

where b is a vector and A is a matrix of proper size. It is easy to see that $\mathcal{P}$ is convex. Let us consider two points $x \in \mathcal{P}$ and $y \in \mathcal{P}$, then, we have the following results:

$$Ax \leq b \tag{3.15}$$
$$Ay \leq b \tag{3.16}$$

Now, let us define an intermediate point $z = \lambda x + (1 - \lambda)y$ with $0 \leq \lambda \leq 1$, then,

$$\lambda Ax + (1 - \lambda)Ay \leq \lambda b + (1 - \lambda)b \tag{3.17}$$
$$Az \leq b \tag{3.18}$$

Therefore, $z \in \mathcal{P}$ which means that $\mathcal{P}$ is convex.

Example 3.5. The following set forms a polytope, as depicted in Figure 3.4:

$$2x + 5y \leq 10$$
$$2y - 2x \leq 1$$
$$x \geq 0 \tag{3.19}$$
$$y \geq 0$$

This is a convex set that is represented as Equation (3.14) with the following parameters:

$$A = \begin{pmatrix} 2 & 5 \\ 2 & -2 \\ -1 & 0 \\ 0 & -1 \end{pmatrix} \tag{3.20}$$

and,

$$b = (10\ 1\ 0\ 0)^{\mathsf{T}} \tag{3.21}$$

Example 3.6. A unit ball with center in zero is a set $\mathcal{B}_0$, defined as follows:

$$\mathcal{B}_0 = \{x \in \mathbb{R}^n : \|x\| \leq 1\} \tag{3.22}$$

where $\|\cdot\|$ represents a norm in $\mathbb{R}^n$. Let us consider two points $x \in \mathbb{B}_0$ and $y \in \mathbb{B}_0$, then we have that,

$$\|x\| \leq 1 \tag{3.23}$$
$$\|y\| \leq 1 \tag{3.24}$$

Figure 3.4 Example of a polytope in the plane.

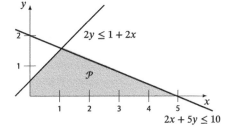

Now, let us consider an intermediate point $z = \lambda x + (1 - \lambda)y$ with $0 \leq \lambda \leq 1$, then we have the following result[1]:

$$\|z\| = \|\lambda x + (1 - \lambda)y\| \tag{3.25}$$

$$\leq \lambda \|x\| + (1 - \lambda)\|y\| \tag{3.26}$$

$$\leq b \tag{3.27}$$

Therefore $z \in \mathbb{B}_0$ and consequently, $\mathbb{B}_0$ is a convex set. The set Ω_A as shown in Figure 3.2) is a particular case of a unitary ball in $\mathbb{R}^2$. Notice that the set $\mathcal{M}_r = \{x \in \mathbb{R}^n : \|x\| \geq r\}$ is not a convex set, for the same reasons exposed in Example 3.1.

In summary, the procedure to demonstrate convexity is straightforward: first, we define two points in the set; then, we define an intermediate point; finally, we demonstrate that this intermediate point belongs to the set. Let us see a final example.

Example 3.7. An ellipsoid is a set given by

$$\mathcal{E} = \{x \in \mathbb{R}^n : x^\mathsf{T} A x \leq 1\} \tag{3.28}$$

where A is a symmetric positive definite matrix (i.e., all its eigenvalues are positive). A set of this form is equivalent to a unit ball, as follows:

$$\mathcal{B} = \{x \in \mathbb{R}^n : \|x\|_A \leq 1\} \tag{3.29}$$

with a norm $\|\cdot\|_A$ defined as Equation (3.30),

$$\|x\|_A = \sqrt{x^\mathsf{T} A x} \tag{3.30}$$

We already demonstrated in the previous example that a unit ball is a convex set. The reader is invited to demonstrate that Equation (3.30) is actually a norm, if A is positive definite.

Example 3.8. A second-order cone (SOC) is a convex set defined as follows:

$$\text{SOC} = \{x \in \mathbb{R}^n : \|Ax + b\| \leq c^\mathsf{T} x + d\} \tag{3.31}$$

where $\|\cdot\|$ represents the 2-norm; A is a square matrix; b and c are vectors in $\mathbb{R}^n$; and d is a scalar. Second-order cones play a crucial role in the convex approximations for the optimal power flow, as presented in Chapter 10.

---

1 We used the properties of the norm, already explained in the previous chapter.

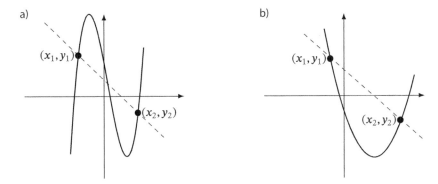

Figure 3.5 Comparison between a convex and a non-convex function.

3.2 Convex functions

As discussed in Chapter 2, an optimization problem is constituted by a feasible set and an objective function. We already observed that the feasible set might be convex. Now, we can extend this concept to the objective function.

A function $f : \mathbb{R}^n \rightarrow \mathbb{R}$ is convex if fulfills the following inequality, commonly known as Jensen's inequality, for any pair of points $x, y \in \mathbb{R}^n$:

$$f(\lambda x + (1 - \lambda)y) \le \lambda f(x) + (1 - \lambda)f(y) \tag{3.32}$$

where $0 \le \lambda \le 1$. If the inequality fulfills strictly, then the function is *strictly convex*[2]. We say the function $f(x)$ is concave if $-f(x)$ is convex.

Figure 3.5 depicts a convex and a non-convex function. In case a) we can draw a line between (x_1, y_1) and (x_2, y_2) but there are some parts of the function that are above the line segment (i.e., the function is non-convex). In case b) we can see that every point in the line segment is below the function itself. This function is convex.

Example 3.9. A list of convex functions in $\mathbb{R}$ is presented in Figure 3.6. A convex function is continuous but not necessarily derivable; for example, $f(x) = |x|$ is a convex function but it does not have derivative at $x = 0$. In addition, a function might be convex only in certain domain; for instance, $f(x) = -\cos(x)$ is convex, only for $x \in [-\pi/2, \pi/2]$.

There are three properties that are useful for defining and identifying convex functions, especially in the general case of $\mathbb{R}^n$.

---

2 This type of functions ensures uniqueness of the solution as will be presented in Section 3.4.

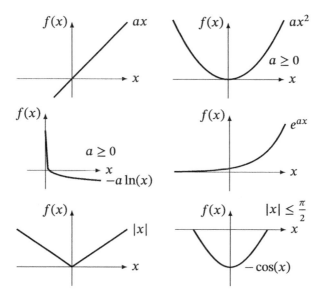

Figure 3.6 List of common convex functions.

Property 1: The sum of two or more convex functions is also a convex function. However, the rest of convex functions is not necessarily convex.

Property 2: The composition of a convex and an affine function is also a convex function, e.g., $f(Ax + b)$ is convex if f is convex.

Property 3: The composition of a convex function with a convex non-decreasing function is also a convex function (see Figure 3.7).

Let us demonstrate the third property (the reader is invited to demonstrate the first two properties). Consider a convex function $f : \mathbb{R}^n \to \mathbb{R}$, then it fulfills Jensen's inequality:

$$f(\lambda x + (1 - \lambda)y) \leq \lambda f(x) + (1 - \lambda)f(y) \tag{3.33}$$

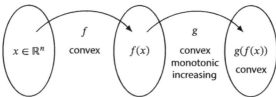

Figure 3.7 Composition of a convex function and a convex non-decreasing function.

Let us consider now, a function g which is monotonic increasing, then we can apply this function to Equation (3.33) without modifying the direction of the inequality:

$$g(f(\lambda x + (1 - \lambda)y)) \leq g(\lambda f(x) + (1 - \lambda)f(y)) \tag{3.34}$$

If g is also convex, then it is easy to see from Equation (3.34) that $g(f(x))$ fulfills also the Jensen's inequality and therefore, it is convex.

Example 3.10. The following functions are convex since they can be defined as the sum of convex functions:

$$f(x) = 3x + 2e^{-x} \tag{3.35}$$

$$f(x) = 5x + 3x^2 \tag{3.36}$$

$$f(x) = 8x^2 - \ln(x) \tag{3.37}$$

Notice that the rest of convex functions is not necessarily convex. For example $f(x) = 5x - 3x^2$ is non-convex.

Example 3.11. The following functions are convex since they are the composition of a convex and an affine function:

$$f(x) = \|a^\top x + b\| \tag{3.38}$$

$$f(x) = (a^\top x + b)^2 \tag{3.39}$$

$$f(x) = -\ln(a^\top x + b) \tag{3.40}$$

Example 3.12. The following function is convex since it is the composition of a convex function $x^\top x$, and a convex/monotonic-increasing function $\exp(x)$:

$$f(x) = \exp(x^\top x) \tag{3.41}$$

3.3 Convex optimization problems

A convex optimization problem is a problem given by Equation (3.42),

$$\min f(x)$$

$$x \in \Omega \tag{3.42}$$

where f is a convex function and Ω is a convex set. The latter can be represented in terms of equality and inequality constraints due to the properties already defined in the previous section.

Let us consider a function $q : \mathbb{R}^n \to \mathbb{R}$ and define a new set known as the epigraph of g, as follows:

$$\text{epi}(g, t) = \{x \in \mathbb{R}^n : g(x) \le t\} \tag{3.43}$$

A function is convex if and only if its epigraph is convex. This property is very useful to identify convex sets and convex functions. For instance, the set Ω_A in Example 3.1 is convex since the function $f(x, y) = x^2 + y^2$ is a convex function.

Example 3.13. An optimization problem of the form

$$\min \ f(x) \tag{3.44}$$

can be transformed using the epigraph into the following problem:

$$\min t$$
$$f(x) \le t \tag{3.45}$$

This model is convex if f is convex.

In general, a set of equations of the form $g_i(x) \le 0$ can be represented as the intersections of the epigraphs of g_i with $t = 0$, namely:

$$\Omega = \bigcap_i \text{epi}(g_i, 0) \tag{3.46}$$

Recall that the intersection of convex sets is also a convex set. Equality constraints of the form $Ax = b$ (affine) may complement the model. Hence, the canonical representation of a convex optimization problem is presented below:

$$\min f(x)$$
$$g(x) \le 0 \tag{3.47}$$
$$Ax = b$$

where f and g are convex functions. It is important to notice that equality constraints must be affine and inequality constraints must be convex. A constraint of the form $f(x) = 0$ or $f(x) \ge 0$ does not define a convex set, even if f is convex (recall the case of Ω_B and Ω_C in Figure 3.2). Below we present some common examples of convex optimization problems.

Example 3.14. A linear programming problem is a problem where both the objective function and the constraints are affine.

$$\min c^T x$$
$$Ax = b \qquad \qquad (3.48)$$
$$x \geq 0$$

This type of problem is convex since affine functions are also convex. The feasible set is a polytope and the optimum is usually a vertex. Several problems in power systems operation can be represented a linear programming models, for example, the economic dispatch and the demand-side management.

Example 3.15. Quadratic programming problems are problems where the objective function is quadratic and the constraints are affine.

$$\min \frac{1}{2} x^T H x + c^T x$$
$$Ax = b \qquad \qquad (3.49)$$
$$x \geq 0$$

This problem is convex if H positive semidefinite. The feasible set is also a polytope but the optimum could be inside the set, or in the boundary. The economic dispatch of thermal units is a common example of quadratic programming problems.

Example 3.16. Quadratic programming with quadratic constrains are problems with the following structure:

$$\min \frac{1}{2} x^T H x + c^T x$$
$$Ax \leq b \qquad \qquad (3.50)$$
$$\frac{1}{2} x^T M x + s^T x \leq t$$

This problem is convex if both H and M are positive semidefinite. It is very important to emphasize in direction of the sign $\leq$ in Equation (3.50). Notice that the following constraints are not convex even if M is semidefinite:

$$\frac{1}{2} x^T M x + s^T x = t \qquad \qquad (3.51)$$

$$\frac{1}{2} x^T M x + s^T x \geq t \qquad \qquad (3.52)$$

Quadratic constraints can be transformed in second-order cones constrains as will be presented in Chapter 5 and in Chapter 10 for the optimal power flow problem.

Example 3.17. A problem with the following objective function is convex, since $\exp(x)$ is a convex function.

$$\min \sum_k \exp(a_k x_k)$$

$$Ax \le b \tag{3.53}$$

Environmental dispatch is an instance of this type of problem.

3.4 Global optimum and uniqueness of the solution

The main feature of convex optimization problems is the capability to guarantee global optimal, therein lies our interest in this type of models. The intuition behind this feature is quite simple: let us consider a convex optimization problem represented as Equation (3.47) with a local optimum $\tilde{x}$, that is to say:

$$f(\tilde{x}) = \inf\{f(x) : x \in \mathcal{N}\} \tag{3.54}$$

where $\mathcal{N} \subset \Omega$ in a given neighborhood is defined as follows:

$$\mathcal{N} = \{x \in \Omega, \|x - \tilde{x}\| < r\} \tag{3.55}$$

Let us consider now, a feasible point $y \in \Omega$ outside of $\mathcal{N}$. Since Ω is convex, we have that:

$$x = \alpha \tilde{x} + \beta y \tag{3.56}$$

where $\alpha + \beta = 1$ y $0 \le \alpha \le 1, 0 \le \beta \le 1$. Now, since f is convex, then we have the following:

$$f(x) \le \alpha f(\tilde{x}) + \beta f(y) \tag{3.57}$$

Evidently $f(\tilde{x}) \le f(x)$ since x is in the neighborhood of $\tilde{x}$, therefore:

$$f(\tilde{x}) \le f(x) \le \alpha f(\tilde{x}) + \beta f(y) \tag{3.58}$$

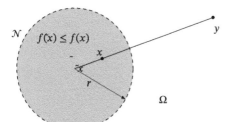

Figure 3.8 Schematic representation of the neighborhood of $\tilde{x}$ in a convex optimization problem over a set Ω.

$$(1 - \alpha)f(\tilde{x}) \leq \beta f(y) \tag{3.59}$$

$$\beta f(\tilde{x}) \leq \beta f(y) \tag{3.60}$$

$$f(\tilde{x}) \leq f(y) \tag{3.61}$$

Consequently, any feasible point outside the neighborhood of $\tilde{x}$ is also greater than $f(\tilde{x})$ and hence, $\tilde{x}$ is the optimal point in the entire set Ω, that is to say, $\tilde{x}$ is a global optimum of the problem.

Now, a global optimum does not imply uniqueness, since we could have several $\tilde{x}$ with the same value of the objective function[3]. We require the function to be strictly convex in order to guarantee the solution is unique. Let us see why this is so: consider two optimal points $\tilde{x}, \tilde{y}$, namely:

$$f(\tilde{x}) = f(\tilde{y}) \leq f(z) \tag{3.62}$$

for all $z \in \Omega$. Let us consider now, an intermediate point $z = \alpha\tilde{x} + \beta\tilde{y}$; since Ω is convex, then this point must belong to the feasible set. If the function is strictly convex, then we have that:

$$f(z) < \alpha f(\tilde{x}) + \beta f(\tilde{y}) = f(\tilde{x}) = f(\tilde{y}) \tag{3.63}$$

but $f(z) < f(\tilde{x})$ contradicts our initial assumption. Therefore, there is no way that $\tilde{x} \neq \tilde{y}$ for a strictly convex function (i.e., $\tilde{x} = \tilde{y}$ and optimum is unique).

One way to identify a strictly convex function is by defining a new function g as follows:

$$g(x) = f(x) - \mu \|x\|^2 \tag{3.64}$$

with $\mu > 0$. If g is convex, then f is not only convex but also strictly convex. The relation among strongly, strictly, and convex functions is depicted in Figure 3.9:

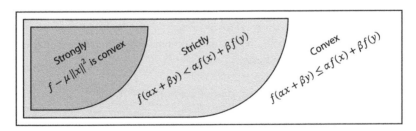

Figure 3.9 Relation among the concepts of strict and strong convexity. All strongly convex functions are strictly convex. Both are, of course, convex.

---

3 see Figure 2.5 in Chapter 2.

3.5 Duality

In this section, we briefly study duality theory for convex optimization problems. We can associate a complementary optimization problem called a dual problem for any optimization problem, convex or not convex. Let us consider a convex optimization problem as Equation (3.47), then we can define a Lagrangian function as follows:

$$\mathcal{L}(x, \lambda, \mu) = f(x) + \mu^\top g(x) + \lambda^\top (Ax - b) \tag{3.65}$$

Where λ are the Lagrange multipliers associated to equality constraints, and μ are the Lagrange multipliers associated to inequality constraints. These multipliers represent a change in the objective function for a change in the constraint, as previously demonstrated in Section 2.6. We assume that $\mu \geq 0$.

Now, we define a new function called the dual function as presented below:

$$W(\lambda, \mu) = \inf_x \mathcal{L}(x, \lambda, \mu) \tag{3.66}$$

Notice the dual function depends on λ and μ but not on x as is the case of $\mathcal{L}$. Since $\mu \geq 0$ then for any feasible point x, we have that,

$$\mathcal{L}(x, \lambda, \mu) \leq f(x) \tag{3.67}$$

Since the point is feasible, then $Ax - b = 0$ and $g(x) \leq 0$, therefore the term $\mu^\top g(x)$ is negative and Equation (3.67) is clearly met. This property is known as *weak duality* and is general for any feasible point, even in the minimum. Therefore, we have also that

$$W(\lambda, \mu) \leq f(x) \tag{3.68}$$

Hence, the dual function is a lower limit of the optimization problem Equation (3.47) (henceforth, the primal problem). We can define a dual optimization problem as given below:

$$\max \ W(\lambda, \mu)$$
$$\mu \geq 0 \tag{3.69}$$

The relation between the primal and the dual problem is depicted in Figure 3.10. The Primal problem search for a minimum point in the feasible set while the dual problem search for a maximum point in its own feasible set. The maximum point of $\mathcal{D}$ constitutes a low boundary of the problem since any

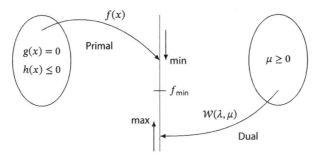

Figure 3.10 Relation between primal and dual problems.

feasible point in the Primal returns a higher value than the value given by any feasible point at the dual problem.

Example 3.18. The dual of a linear programming problem is calculated as follows:

$$W(\mu) = \inf_{x} \{c^\mathsf{T} x + \mu^\mathsf{T}(Ax - b)\} \tag{3.70}$$

This infimum exists if $(c^\mathsf{T} + \mu^\mathsf{T} A) = 0^\mathsf{T}$, and thus, $W(\mu) = -\mu^\mathsf{T} b$. Therefore, the dual problem is:

$$\min b^\mathsf{T} \mu$$
$$c + A^\mathsf{T} \mu = 0 \tag{3.71}$$
$$\mu \geq 0$$

That is to say, the dual of a linear programming problem is another linear programming problem.

Example 3.19. Let us consider the following quadratic programming problem, that represents a simple economic dispatch:

$$\min \sum_{k=1}^{n} \frac{a_k}{2} p_k^2 + b_k p_k$$
$$\sum_{k=1}^{m} p_k = d \tag{3.72}$$
$$p_k \geq 0$$

The dual function is given below:

$$W(\mu, \lambda) = \inf_p \left\{ \sum_{k=1}^n \left(\frac{a_k}{2} p_k^2 + b_k p_k \right) + \lambda \left(d - \sum_{k=1}^m p_k \right) + \sum_{k=1}^m \mu_k(-p_k) \right\}$$

$$(3.73)$$

and hence, the dual problem is the following:

$$\max \lambda d - \sum_{k=1}^n \frac{1}{2a_k} (\lambda + \mu_k - b_k)^2$$

$$\mu_k \geq 0 \qquad\qquad\qquad (3.74)$$

In general, the dual of a quadratic programming problem is another quadratic programming problem.

The dual function is concave, even if the Primal problem is non-convex. Remember that a function W is concave if $-W$ is convex. Therefore, we must check if $-W$ holds Jensen's inequality. Let us take two points in the dual function, (λ_1, μ_1) and (λ_2, μ_2) with two real values $\alpha \geq 0$, $\beta \geq 0$ with $\alpha = 1 - \beta$; then, let us evaluate the function in a generic midpoint, as presented below [4]:

$$-W(\alpha\lambda_1 + \beta\lambda_2, \alpha\mu_1 + \beta\mu_2) = -\inf_x \{\mathcal{L}(x, \alpha\lambda_1 + \beta\lambda_2, \alpha\mu_1 + \beta\mu_2)\}$$

$$= \sup_x \{-\mathcal{L}(x, \alpha\lambda_1 + \beta\lambda_2, \alpha\mu_1 + \beta\mu_2)\}$$

$$\leq \alpha\sup_x \{-\mathcal{L}(x, \lambda_1, \mu_1)\} + \beta\sup_x \{-\mathcal{L}(x, \lambda_2, \mu_2)\}$$

$$= -\alpha W(\lambda_1, \mu_1) - \beta W(\lambda_2, \mu_2) \qquad (3.75)$$

Therefore, W is concave, and $\max W = -\min -W$ is a convex problem which is, sometimes, easier to solve than the primal problem.

Example 3.20. Let us consider the following optimization problem:

$$\min f(x) = x^4 - 5x^2 - 3x$$

$$x \geq 1 \qquad\qquad\qquad (3.76)$$

The objective function of this problem is the polynomial plotted in Figure 3.11a which is, evidently, non-convex. However, the dual function is convex and defines a lower bound of the problem as shown in Figure 3.11b.

---

4 Recall that $\inf(W) = -\sup(-W)$.

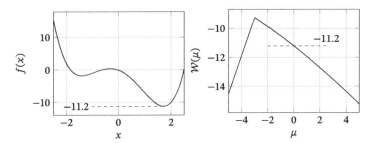

Figure 3.11 Example of a primal non-convex problem and its corresponding dual function.

Weak duality allows finding a lower limit of the primal problem by maximizing the dual. However, very often, the dual is equal to the primal. In that case, we say the problem fulfills the *strong duality conditions*. There is a simple criterion for strong duality in convex optimization problems. Informally, the primary condition to guarantee strong duality is that the feasible region must have a non-empty relative interior. This criterion is named as Slater's conditions [17].

Let us consider the feasible set Ω, of a primal convex optimization problem, namely:

$$\Omega = \{x \in \mathbb{R}^n : Ax - b = 0, \ g(x) \leq 0\} \tag{3.77}$$

we define a new set known as the relative interior relint(Ω) as follows:

$$\text{relint}(\Omega) = \{x \in \mathbb{R}^n : Ax - b = 0, \ g(x) < 0\} \tag{3.78}$$

Notice the only difference between Ω and relint(Ω) is in the inequality constraint. Slater's condition state that relint(Ω) $\neq \emptyset$, that is to say, there is at least one point that fulfills equality and the inequality constraints strictly. Let us analyze these concepts in the following example:

Example 3.21. Consider the following optimization problem

$$\min x^2$$

$$x \geq 3 \tag{3.79}$$

The Lagrangian function is

$$\mathcal{L}(x, \mu) = x^2 + \mu(3 - x) \tag{3.80}$$

There is no λ because the model has only inequality constraints. The dual function is calculated taking the minimum of this function (i.e., when $2x - \mu = 0$):

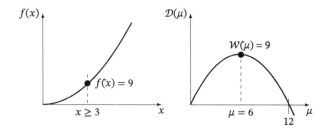

Figure 3.12 Comparison between the primal and the dual problems.

$$W(\mu) = \inf_{x} \mathcal{L}(x, \mu) = 3\mu - \mu^2/4 \tag{3.81}$$

Now, we can define a new optimization problem as max $W(\mu)$ with $\mu \geq 0$. Both, the primal and the dual problems are shown in Figure 3.12; as expected, any feasible point in W is below that any feasible point in f (feasible points in the primal are those $x \geq 3$ and feasible points in the dual are those $\mu \geq 0$). In addition, there are feasible points that fulfills strictly the inequality constraint (for example $x = 4$), therefore, we can guarantee that min $f(x) = $ max $W(\mu)$ (in this case, the optimum is $f(\tilde{x}) = 9$).

Dual variables represent the change in the objective function for a change in the constrain. Let us consider an optimization problem as

$$\min f(x)$$
$$Ax = b \tag{3.82}$$
$$g(x) \leq t$$

Let us suppose we know the optimum $\tilde{x}$ and the corresponding dual variables $\tilde{\lambda}$ and $\tilde{\mu}$, then we have that:

$$\tilde{\lambda} = \frac{\Delta f(\tilde{x})}{\Delta b} \tag{3.83}$$

$$\tilde{\mu} = \frac{\Delta f(\tilde{x})}{\Delta t} \tag{3.84}$$

This interpretation of the dual variables is vital in common practical problems. These variables can be used to define new investments, as explained in Chapter 7.

3.6 Further readings

A summary of the main properties of convex optimization problems is presented in Table 3.1. A problem is convex if the objective function and inequality constraints are convex, and the equality constraints are affine. Once we have checked these conditions, we can conclude that the optimum is global. Therein lies the importance of identifying convex functions in an optimization model. In this chapter, we presented common examples of convex functions. More exotic functions can be studied in [17]. Another type of convex functions such as second-order cone (SOC) and semidefinite programming (SDP) will be studied in Chapter 5. Uniqueness can be guaranteed by using the concepts of strong and strict convexity; usually, it is simpler to identify a strongly convex function. More details about these types of functions can be studied in [18]. The concept of duality can be studied in more detail in [19]. We only presented the most basic concept of duality and some conditions to interpret results. The interpretation of the dual variables is critical in many practical problems. Other concepts such as the Karush–Kuhn–Tucker conditions can be studied in [20]. Finally, it is recommended to review general concepts of linear algebra; two excellent references are [21] and [16]. We avoided solving optimization problems by hand since our objective is to solve large optimization problems. All repetitive calculations must be made by the computer, as presented in the next chapter.

Table 3.1 Summary of the main properties of convex optimization problems.

Definition	Consequence
Convex problem	
min $f(x)$, convex	global optimum
$Ax = b$, affine	
$g(x) \leq 0$, convex	
Strictly convex function	
$f(\alpha x + \beta y) < \alpha f(x) + \beta f(y)$	unique solution
Strongly convex function	
$f(x) - \mu \|x\|^2$ also convex	strictly convex
Dual function	
$\mathcal{W}(\mu, \lambda) = \inf \left\{ f(x) + \lambda^{\top}(Ax - b) + \mu^{\top}g(x) \right\}$	concave
Dual problem	
max $\mathcal{W}(\mu, \lambda)$ with $\mu \geq 0$	weak duality
	max $\mathcal{W} \leq \min f$
Slater conditions	
There exists at least one x such that	strong duality
$Ax + b = 0$, and $g(x) < 0$	max $\mathcal{W} = \min f$
Dual variables	change of the objective function
λ, μ	for a change in the constraint

3.7 Exercises

1. Show that the following set is convex:

$$\Omega = \{(x, y) \in \mathbb{R}^2 : x \geq 0, y \geq 0, xy \geq 1\} \tag{3.85}$$

2. Show that the sum of two convex functions is also a convex function.
3. Show that the composition of a convex function and an affine function, results in an convex function.
4. Demonstrate that a function is convex if and only if its epigraph is convex.
5. Identify which of these functions are convex:

$$f(x) = x^T A x, \text{ with } A > 0 \tag{3.86}$$

$$f(x) = \exp((Bx + c)^T A(Bx + c)), \text{ with } A \geq 0 \tag{3.87}$$

$$f(x) = -\ln((Bx + c)^T A(Bx + c)), \text{ with } A \geq 0 \tag{3.88}$$

6. Show the following relation using Jensen's inequality (use the fact that $-\ln(x)$ is convex and monotone).

$$\prod_{k=1}^{n} x_k^{1/n} \leq \sum_{k=1}^{n} \frac{1}{n} x_k \tag{3.89}$$

7. Show the Lagrangian and the dual function associated to the following optimization problem:

$$\min f(x) = (x - 1)^2$$
$$x^2 \leq 0 \tag{3.90}$$

Show the feasible space of both the dual and the primal problems. Analyze the conditions of weak and strong duality.

8. Determine the dual problem associated to a quadratically constrained quadratic programme presented below:

$$\min \frac{1}{2} x^T H x + bx$$
$$\frac{1}{2} x^T A x + cx \leq d \tag{3.91}$$

9. Consider the following optimization problem

$$\min x^2 + 1$$
$$(x - 2)(x - 4) \leq 0 \tag{3.92}$$

Show the set of feasible solutions and find the optimum. Plot the objective function vs x; on the same graph, plot the Lagrangian function $\mathcal{L}(x, \mu)$ for different values of μ (for example $\mu = 1, \mu = 5$, and $\mu = 10$). Formulate the dual problem and solve it. Analyze the conditions of strong and weak duality.

10. Solve the following problem using both the primal and the dual formulations (use graphs of the function and the feasible set in order to find the optimum)

$$\min x + y$$
$$x^2 + y^2 \leq 1 \tag{3.93}$$

4

Convex Programming in Python

Learning outcomes
By the end of this chapter, the student will be able to: • Identify linear and quadratic optimization problems.   • Solve linear and quadratic problems using Python.

4.1 Python for convex optimization

In the previous chapter, we learned that local optima are guaranteed to be global in convex optimization problems. Other theoretical properties, such as the uniqueness of the solution and strong duality, were also assured under well-defined conditions. These properties are intrinsic of the model, regardless of the solution algorithm, and hence, any solver will give the same solution.

We use Python and the module CvxPy, as a modeling language for convex optimization problems [10]. This module checks the convexity of the problem and calls a solver that returns the solution of the model (see Figure 4.1); then, results can be analyzed in Python using other modules, such as NumPy for linear algebra operations, and MatplotLib for plotting the results. In this way, Python is transformed into a complete modeling language that permits writing and analyzing the optimization problem in a systematic form. However, we must carefully define the model to ensure it is convex, using a set of functions and rules for constructing the model and guarantee convexity. This philosophy for solving optimization problems is known as disciplined convex programming [22].

Mathematical Programming for Power Systems Operation: From Theory to Applications in Python. First Edition. Alejandro Garcés.
© 2022 by The Institute of Electrical and Electronics Engineers, Inc. Published 2022 by John Wiley & Sons, Inc.

Figure 4.1 Using Python for mathematical optimization.

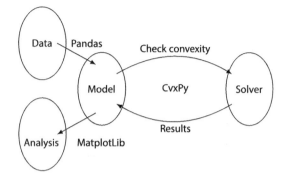

Python and CvxPy allow us to concentrate on the model without worrying about the solution algorithm; since, as mentioned above, the model is well-behaved when it is convex. The following sections present the implementation of linear and quadratic programming models to solve general optimization problems. After that, a brief review of the algorithms is presented in order to understand what is inside the box.

4.2   Linear programming

Reality is non-linear in nature. However, in many cases, we can formulate linear programming approximations that simplify complicated problems. In a linear programming problem, both the objective function and the constraints are affine[1], generating a polytope as the feasible set. The canonical representation of a linear programming problem is presented below:

$$\min c^\mathsf{T} x$$
$$Ax = b \tag{4.1}$$
$$x \geq 0$$

Where $x \in \mathbb{R}^n$, are decision variables; $c \in \mathbb{R}^n$ is a vector that usually represents costs; $b \in \mathbb{R}^m$ is a vector that represents resources; and, $A \in \mathbb{R}^{m \times n}$ is a matrix that defines physical constraints of the process that is being optimized. Any linear programming problem can be transformed into the canonical representation. However, this is not the only representation, and sometimes, it is not the most convenient either. Practical applications will come with their own representation.

---

1 Remember that an affine function is of the form $f(x) = ax + b$.

Example 4.1. Let us transform the following linear programming problem, to the canonical form:

$$\max 3x - 2y$$
$$x + y \leq 5 \tag{4.2}$$
$$x, y \geq 0$$

First, we multiply objective function by -1 to obtain a minimization problem; then, we define a slack variable z that transforms the inequality into equality; finally, we organize the model as follows:

$$\min \ -3x + 2y$$
$$x + y + z = 5 \tag{4.3}$$
$$x, y, z \geq 0$$

In this case, $c = (-3, 2)^{\mathsf{T}}$, $A = (1, 1, 1)$, and $b = 5$.

The set of feasible solutions of a linear programming problem is a geometric object known as polytope[2]. The optimum of a linear programming problem, if it exists, is placed in a vertex. Therefore, optimization algorithms, such as the simplex method, search among the vertices until it achieves the optimum. For problems in $\mathbb{R}^2$ and $\mathbb{R}^3$, we can use the gradient direction to identify the optimal direction and find graphically the optimal solution. The next example shows the methodology.

Example 4.2. Let us consider the following linear programming problem:

$$\max 3x_0 + 3x_1$$
$$x_0 + 2x_1 \leq 4$$
$$4x_0 + 2x_1 \leq 12$$
$$-x_0 + x_1 \leq 1 \tag{4.4}$$
$$x_0 \geq 0$$
$$x_0 \geq 0$$

The set of feasible solutions for this problem is shown in Figure 4.2. The direction of the gradient of the objective function is represented as ∇obj; using this direction, we can easily identify the optimum. As expected, the optimum is in a vertex, in this case in $d = (8/3, 2/3)$.

---

2 We already studied this object in Example 3.4 (Chapter 3) and concluded that it is convex.

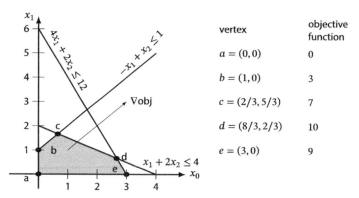

Figure 4.2 Set of feasible solutions for the linear programming problem in Example 4.2.

Practical linear programming problems for power systems operation could have thousands of decision variables and constraints. Therefore, the graphical method is not enough; we require to solve these problems using a computer. In the following examples, we show how to use Python and CvxPy to solve linear programming problems.

Example 4.3. Let us solve the linear programming problem presented in Example 4.1 using Python. The script that solves this problem is presented below:

```python
import numpy as np
import cvxpy as cvx
x = cvx.Variable()
y = cvx.Variable()
obj = cvx.Maximize(3*x-2*y)
res = [x+y <= 5, x>=0, y>=0]
Model = cvx.Problem(obj,res)
Model.solve()
print(np.round(obj.value),np.round(x.value,2),
np.round(y.value,2))
```

The code is intuitive; it starts by defining decision variables with the command `cvx.Variable`; then, the objective function and the set of constraints are determined; the set of constraints are stored in a vector named `res`; after that, the model is solved, and results are printed, rounded to two decimal places. Notice that we did not require to change the problem to a canonical form; instead, the problem was represented as was raised.

We can use different solvers and see the iterations on each solver as shown below. The complete list of solvers is available in [23].

```
Model.solve(solver=cvx.OSQP,verbose=True)
print(obj.value,x.value,y.value)
Model.solve(solver=cvx.ECOS,verbose=True)
print(obj.value,x.value,y.value)
Model.solve(solver=cvx.SCS,verbose=True)
print(obj.value,x.value,y.value)
```

Example 4.4. Let us solve the linear programming problem presented in Example 4.2.

```
import cvxpy as cvx
x = cvx.Variable(2,nonneg=True)
obj = cvx.Maximize(3*x[0]+3*x[1])
res = [  x[0] + 2*x[1]  <= 4,
       4*x[0] + 2*x[1]  <= 12,
        -x[0] +   x[1]  <= 1]
Model = cvx.Problem(obj,res)
Model.solve(solver=cvx.SCS)
print('objective:',obj.value)
print('decisions:',x.value)
```

The script is quite similar to the previous example. The only difference was in the definition of the variables x, which is a vector in $\mathbb{R}^2_+$. Therefore, we defined the size of the variable and a condition no no-negativity (i.e., nonneg= True).

Example 4.5. The transportation problem is a special type of linear programming problem, which consists in minimizing the cost of transporting a commodity, from a set of sources to a set of destinations (this commodity can be, off course, electric power). Each source has a limited supply while each destination has a demand to be satisfied. Decision variables are represented in a matrix x, where x_{ij} represents the amount of products transported from i to j. Each route has unit costs c_{ij} and the amount of products available in the sources is represented by s_i, while the amount of product demanded in each destination is d_j. The problem consists on minimizing total costs, constrained to the balance of each source and destination. A general mathematical model is presented below:

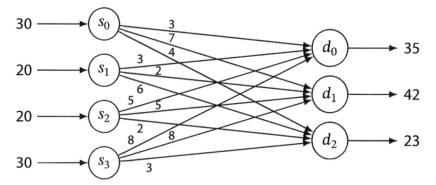

Figure 4.3 Oriented graph for a transportation problem with four sources (*s*) and three destinations (*d*). The numbers in the arrows corresponds to the unit costs for each route.

$$\min \sum_{i}^{m} \sum_{j}^{n} c_{ij} x_{ij}$$

$$\sum_{j}^{n} x_{ij} = s_i$$

$$\sum_{i}^{m} x_{ij} = d_j \qquad\qquad (4.5)$$

$$x_{ij} \geq 0$$

Let us consider a problem with $m = 4$ sources and $n = 3$ destinations, as shown in Figure 4.3. All the parameters required to solve the problem are depicted in the figure.

The implementation in Python for this transportation problem is given below. The reader is invited to evaluate the solution and analyze the solution.

```python
import cvxpy as cvx
import numpy as np
m = 4
n = 3
c = np.array([[3,3,5,8],[7,2,5,8],[4,6,2,3]])
s = np.array([30,20,20,30])
d = np.array([35,42,23])
x = cvx.Variable((m,n),nonneg=True)
obj = cvx.Minimize(cvx.trace(c@x))
res = [x@np.ones(n)    == s,
          x.T@np.ones(m) == d]
Model = cvx.Problem(obj,res)
```

```
Model.solve()
print(np.round(x.value,2))
```

Again, we avoid constraints in the form $x \geq 0$, by defining the variable as positive, with the modifier `nonneg=True`. Notice that equality constraints were defined by `==`.

4.3 Quadratic forms

It is common to find optimization problems in power systems operations that have quadratic objective functions. For example, the classic economic dispatch of thermal units is usually a quadratic problem. Therefore, it is essential to understand some properties of these types of functions.

A quadratic form $q : \mathbb{R}^n \to \mathbb{R}$ is a multivariate polynomial with variables $x_0, x_1, ...x_{n-1}$, where all the terms are, at most, of order 2. Any quadratic form can be written as given in Equation (4.6):

$$q(x) = x^T A x + b^T x + c \tag{4.6}$$

where A is a square matrix, b is a column vector, and c is a constant.

Example 4.6. Let us consider the following multivariate polynomial:

$$q(x_0, x_1) = 5x_0^2 + 2x_0x_1 + 3x_1^2 + 8x_0 + 4x_1 + 15 \tag{4.7}$$

This polynomial is a quadratic form because the maximum exponent is 2. Therefore, it can be written as Equation (4.6) with the following structure:

$$q(x_0, x_1) = \begin{pmatrix} x_0 \\ x_1 \end{pmatrix}^T \begin{pmatrix} 5 & 2 \\ 0 & 3 \end{pmatrix} \begin{pmatrix} x_0 \\ x_1 \end{pmatrix} + \begin{pmatrix} 8 \\ 4 \end{pmatrix}^T \begin{pmatrix} x_0 \\ x_1 \end{pmatrix} + 15 \tag{4.8}$$

Notice that, this representation is not unique. Another possible representation is presented below:

$$q(x_0, x_1) = \begin{pmatrix} x_0 \\ x_1 \end{pmatrix}^T \begin{pmatrix} 5 & 10 \\ -8 & 3 \end{pmatrix} \begin{pmatrix} x_0 \\ x_1 \end{pmatrix} + \begin{pmatrix} 8 \\ 4 \end{pmatrix}^T \begin{pmatrix} x_0 \\ x_1 \end{pmatrix} + 15 \tag{4.9}$$

It is important to note that not all quadratic forms are convex. In this case, the form is convex since it describes a paraboloid, as may be easily demonstrated by plotting the function.

The main properties of a quadratic form are determined by the matrix A. Two interesting cases are when this matrix is symmetric and when it is skew-symmetric. A matrix A is symmetric when $A = A^T$ and skew-symmetric when $A = -A^T$. Let us analyze the latter case.

Consider a quadratic form $q(x) = x^\mathsf{T} N x$, where N is skew-symmetric; in that case, we have the following[3]:

$$q(x) = q(x)^\mathsf{T} \tag{4.10}$$

$$= (x^\mathsf{T} N x)^\mathsf{T} \tag{4.11}$$

$$= (Nx)^\mathsf{T} x \tag{4.12}$$

$$= x^\mathsf{T} N^\mathsf{T} x \tag{4.13}$$

$$= -x^\mathsf{T} N x = -q(x) \tag{4.14}$$

Therefore, $q(x) = -q(x)$ for any value of x and hence, $q(x) = 0$. In conclusion, a quadratic form $q(x) = x^\mathsf{T} N x$ is zero if N is skew-symmetric.

Any square matrix A can be written in terms of a symmetric and a skew-symmetric matrix, as follows:

$$A = \frac{1}{2}(A + A) + \frac{1}{2}(A^\mathsf{T} - A^\mathsf{T}) \tag{4.15}$$

$$= \frac{1}{2}(A + A^\mathsf{T}) + \frac{1}{2}(A - A^\mathsf{T}) \tag{4.16}$$

$$= \frac{1}{2}M + \frac{1}{2}N \tag{4.17}$$

where $M = A + A^\mathsf{T}$ and $N = A - A^\mathsf{T}$. Matrix M is symmetric (i.e., $M = M^\mathsf{T}$) and N is skew-symmetric (i.e., $N = -N^\mathsf{T}$). Therefore, we have the following:

$$q(x) = x^\mathsf{T} A x \tag{4.18}$$

$$= \frac{1}{2}x^\mathsf{T} M x + \frac{1}{2}x^\mathsf{T} N x \tag{4.19}$$

$$= \frac{1}{2}x^\mathsf{T} M x \tag{4.20}$$

In other words, any quadratic form $q(x)$ can be written in terms of a symmetric matrix.

Example 4.7. The quadratic form given in Equation (4.7) can be written in terms of a symmetric matrix as given below:

$$q(x_0, x_1) = \frac{1}{2} \begin{pmatrix} x_0 \\ x_1 \end{pmatrix}^\mathsf{T} \begin{pmatrix} 10 & 2 \\ 2 & 6 \end{pmatrix} \begin{pmatrix} x_0 \\ x_1 \end{pmatrix} + \begin{pmatrix} 8 \\ 4 \end{pmatrix}^\mathsf{T} \begin{pmatrix} x_0 \\ x_1 \end{pmatrix} + 15 \tag{4.21}$$

---

3 Notice that $q(x)$ is a scalar, not a matrix, and hence $q = q^\mathsf{T}$.

where

$$\begin{pmatrix} 5 & 10 \\ -8 & 3 \end{pmatrix} = \frac{1}{2}\left(\begin{pmatrix} 5 & 10 \\ -8 & 3 \end{pmatrix} + \begin{pmatrix} 5 & -8 \\ 10 & 3 \end{pmatrix}\right) = \frac{1}{2}\begin{pmatrix} 10 & 2 \\ 2 & 6 \end{pmatrix}$$

(4.22)

Any quadratic form $q(x)$ given by Equation (4.6) is continuous and has derivative. Its gradient is given by Equation (4.23):

$$\nabla Q = (A + A^T)x + b$$

(4.23)

since A is symmetric, then $\nabla Q = 2Ax + b$, and its hessian is simply $2A$.

4.4 Semidefinite matrices

There is a particular type of quadratic forms that are always positive. The matrices that represent these forms are known as semidefinite matrices. Thus, we say a symmetric matrix $A \in \mathbb{R}^{n \times n}$ is positive semidefinite if $q(u) = u^T Au \geq 0$ for any $u \in \mathbb{R}^n$. These types of matrices are represented as $A \geq 0$ (notice that the symbol is different from $\geq$). Moreover, we say that the matrix is positive definite $(A > 0)$ if $q(u) > 0$ for any $u \neq 0$. Likewise, we say the matrix is negative definite or semidefinite if $(-A) \geq 0$ or $(-A) > 0$, respectively.

A symmetric and positive semidefinite matrices have the following properties:

- Its eigenvalues are all positive.
- The matrix A can be factorized as $A = CC^T$ where C is a triangular matrix. This is called Cholesky factorization, and the matrix C is usually represented as $A^{1/2}$.
- If $A > 0$ then $A^{-1} > 0$
- If $A > 0$ and $B > 0$ then $A + B > 0$
- However, if $A > 0$ and $B > 0$ we cannot say anything about AB.
- if $A > 0$ and $B > 0$ then $A \circ B > 0$ where $\circ$ represents the Hadamard product (i.e., the point-wise product).

Example 4.8. The following matrix is positive definite

$$A = \begin{pmatrix} 2 & 1 \\ 1 & 1 \end{pmatrix}$$

(4.24)

since its eigenvalues are both positive, i.e., $\lambda = \{0.3819, 2.61803\}$, and have Cholesky factorization[4]:

$$A = \begin{pmatrix} 2 & 1 \\ 1 & 1 \end{pmatrix} = \begin{pmatrix} \sqrt{2} & \sqrt{2}/2 \\ 0 & \sqrt{2}/2 \end{pmatrix}^\mathsf{T} \begin{pmatrix} \sqrt{2} & \sqrt{2}/2 \\ 0 & \sqrt{2}/2 \end{pmatrix} \qquad (4.25)$$

This matrix defines the following quadratic form, which is evidently positive for any $x \neq 0$:

$$q(x) = x^\mathsf{T} A x = 2x_0^2 + 2x_0 x_1 + x_1^2 \qquad (4.26)$$

$$= (x_1 + x_0)^2 + x_0^2 \qquad (4.27)$$

Example 4.9. We can define a function in Python that identify positive definite matrices using the eigenvalues. The code for this function is presented below:

```python
import numpy as np
def IsSD(M):
    Lmin = min(np.linalg.eigvals(M))
    if (Lmin==0):
        print('Positive semidefinite')
    if (Lmin>0):
        print('Positive definite')
    if (Lmin<0):
        print('It is not positive semidefinte')
# usage
A = [[2,1],[1,1]]
IsSD(A)
```

Example 4.10. A quadratic function with a positive semidefinite matrix is convex. Let us consider the following two examples:

$$q_1(x, y) = \frac{1}{2} \begin{pmatrix} x \\ y \end{pmatrix}^\mathsf{T} \begin{pmatrix} 1 & 1/4 \\ 1/4 & 1 \end{pmatrix} \begin{pmatrix} x \\ y \end{pmatrix} \qquad (4.28)$$

$$q_2(x, y) = \frac{1}{2} \begin{pmatrix} x \\ y \end{pmatrix}^\mathsf{T} \begin{pmatrix} 1 & 1/4 \\ 1/4 & -1 \end{pmatrix} \begin{pmatrix} x \\ y \end{pmatrix} \qquad (4.29)$$

---

4 The command in Python for Cholesky factorization is np.linalg.cholesky(A) and the command for calculating the eigenvalues is np.linalg.eigvals(A), where np comes form the NumPy module.

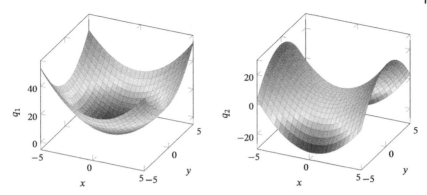

Figure 4.4 Example of two quadratic functions, q_1 (left) is convex whereas q_2(right) is not.

In this case, the eigenvalues of the matrix associated to q_1 are $\lambda = \{1.25, 0.75\}$ whereas the eigenvalues of the matrix associated to q_2 are $\lambda = \{1.03, -1.03\}$. Clearly, q_1 is convex whereas q_2 is not (see Figure 4.4).

4.5 Solving quadratic programming problems

A quadratic form $q(x)$ is convex if its symmetric matrix is positive semidefinite[5]. In that case, we can use CvxPy for solving problems that involves quadratic forms in the objective function, or in the inequality constraints. The first case, is known as quadratic programming and the second case, is known as quadratic programming with quadratic constraints. Let us consider the second case that is more general, namely:

$$\min q_0(x)$$
$$q_1(x) \leq 0 \tag{4.30}$$

where both q_0 and q_1 are convex quadratic forms. It is important to remark that the constraint must be an inequality of the type ≤ 0, otherwise the problem is non-convex. Consider the following optimization problem:

$$\min q_0(x)$$
$$q_1(x) \geq 0 \tag{4.31}$$
$$q_2(x) = 0$$

---

5 Besides, it is strictly convex if it is positive definite.

This problem is not convex, even if q_1 and q_2 were a convex quadratic form, because equality constraints must be affine and inequality constraints must be of the type $\leq$ (see Example 3.1 in Chapter 3). In the following sections, we present a series of simple examples to familiarize ourselves with the functions available in CvxPy, for solving quadratic problems.

Example 4.11. An unconstrained quadratic-convex optimization problem is trivial and can be solved directly. Let us consider the following problem:

$$\min \; q(x) = \frac{1}{2}x^\top H x + b^\top x + c \tag{4.32}$$

where $H = H^\top > 0$ (symmetric and positive definite), then we have the following:

$$\nabla q(x) = Hx + b = 0 \tag{4.33}$$

$$x = -H^{-1}b \tag{4.34}$$

we require that H is positive definite in order to guarantee the inverse exists. In case the matrix is only semi defined, we can use the Moore–Penrose inverse to find a solution for x [16].

Example 4.12. Let us consider the following optimization problem:

$$\min \; 5x_0^2 + 2x_0 x_1 + 3x_1^2 + 7x_0 + x_1 + 10$$

$$x_0 + x_1 = 1 \tag{4.35}$$

$$x_0, x_1 \geq 0$$

Where the objective function can be written as the following quadratic form:

$$q(x_0, x_1) = \begin{pmatrix} x_0 \\ x_1 \end{pmatrix}^\top \begin{pmatrix} 5 & 2 \\ 0 & 3 \end{pmatrix} \begin{pmatrix} x_0 \\ x_1 \end{pmatrix} + \begin{pmatrix} 7 \\ 1 \end{pmatrix}^\top \begin{pmatrix} x_0 \\ x_1 \end{pmatrix} + 10 \tag{4.36}$$

The script in Python for solving this problem is presented below:

```python
import numpy as np
import cvxpy as cvx
A = np.matrix([[5,2],[0,3]])
H = 1/2*(A+A.T)
IsSD(H)
b = np.array([7,1])
c = 10
x = cvx.Variable(2, nonneg = True)
q = cvx.quad_form(x,H)+b.T@x + c
```

```
obj = cvx.Minimize(q)
res = [x[0]+x[1]==1]
Model = cvx.Problem(obj,res)
Model.solve()
print(x.value)
```

First, we define a matrix for the quadratic form and build a symmetric equivalent (see Example 4.7). Then, we determine if it is semidefinite using the function created in Example 4.8; the quadratic form is then represented by the function `quad_form` that is part of the module CvxPy. The rest of the model is intuitive.

Example 4.13. Let us define the unitary ball in $\mathbb{R}^2$ with center in $a = (a_0, a_1)$ as $\mathcal{B}_a$, given by Equation (4.37),

$$\mathcal{B}_a = \{(x_0, x_1) : (x_0 - a_0)^2 + (x_1 - a_1)^2 \leq 1\} \tag{4.37}$$

we are interested in finding a point that minimizes the function $f(x_0, x_1) = x_0 + x_1$, such that (x_0, x_1) belongs to the intersection of the unit balls $\mathcal{B}_a$ and $\mathcal{B}_b$; with $a = (1, 1)$ and $b = (0, 0)$, namely:

$$\min x_0 + x_1$$

$$(x_0, x_1) \in \mathcal{B}_a \cap \mathcal{B}_b \tag{4.38}$$

This problem is convex since unit balls are convex sets. Moreover, each ball can be represented as quadratic forms as presented below:

$$\mathcal{B}_a = \{x \in \mathbb{R}^n : (x - a)^T I (x - a) \leq 1\} \tag{4.39}$$

where I is the identity matrix, and $a \in \mathbb{R}^n$ is a vector that represents the center of the ball. So, a script to solve Model Equation (4.38) is presented below:

```
import numpy as np
import cvxpy as cvx
x = cvx.Variable(2, nonneg = True)
a = np.array([1,1])
q_a = cvx.quad_form(x-a,np.identity(2))
q_b = cvx.quad_form(x,np.identity(2))
obj = cvx.Minimize(x[0]+x[1])
res = [q_a <= 1, q_b <= 1]
Model = cvx.Problem(obj,res)
Model.solve()
print(x.value)
```

The student is invited to plot this problem and analyze the results.

4.6   Complex variables

Several problems in power systems operation have a simple representation in the set of the complex numbers, hence it is natural to formulate optimization problems, using complex decision variables. However, we must be very careful in this type of formulation. Unlike the real and the integer numbers, that are totally ordered set, the complex numbers are not. In general, an optimization model on the complexes may have the following canonical representation, namely:

$$\min f(z)$$
$$g(z) \leq 0$$
$$h(z) = 0 \qquad\qquad (4.40)$$
$$z \in \mathbb{C}^n$$

which is similar to the representation given in Equation (3.47). Nonetheless, we must ensure the co-domain of both the objective function and the inequality constraints, is the real numbers, that is to say $f : \mathbb{C}^n \rightarrow \mathbb{R}$ and $g : \mathbb{C}^n \rightarrow \mathbb{R}$. A constraint in the form $g(x) \leq 0$ does not have sense if the image of g is also complex. Of course Equation (4.40) may be also represented in terms of real and imaginary variables, as given below:

$$\min f(x, y)$$
$$g(x, y) \leq 0$$
$$\text{real}\,(h(x, y)) = 0 \qquad\qquad (4.41)$$
$$\text{imag}\,(h(x, y)) = 0$$
$$x, y \in \mathbb{R}^n$$

where $z = x + jy$. However, (4.40) is a more compact representation of the problem in many practical applications. Convexity and similar mathematical properties must be evaluated in Equation (4.41). Thus, Equation (4.40) is, in most of the cases, only a convenient representation of the problem.

Example 4.14. The optimization problem presented in Equation (4.38) may be written in terms of complex variables; for this, we define a complex variable $z = x_0 + x_1 j$, and the optimization model presented below:

$$\min z_{\text{real}} + z_{\text{imag}}$$
$$\|z - (1 + j)\| \leq 1 \qquad\qquad (4.42)$$
$$\|z\| \leq 1$$

The script for solving this problem is the following:

```
z = cvx.Variable(complex=True)
obj = cvx.Minimize(cvx.real(z)+cvx.imag(z))
res = [cvx.abs(z-(1+1j)) <= 1,
       cvx.abs(z) <= 1]
Model = cvx.Problem(obj,res)
Model.solve()
print(z.value)
```

4.7 What is inside the box?

In this chapter, we used CvxPy as a modeling platform. This module call other solvers to find an optimal solution to the problem. Most of these solvers use efficient variations of the gradient and Newton's methods for unconstrained problems, whereas inequality constrained problems are usually solved by the Interior Point method or similar barrier methods [17]. Details of these methods are beyond the objectives of this book; however, it is interesting to see the main idea by using the following problem:

$$\min f(x)$$
$$g(x) \leq 0 \tag{4.43}$$

we define an indicator function for the inequality constraint as follows:

$$I_g(x) = \begin{cases} 0 & \text{if } g(x) \leq 0 \\ \infty & \text{otherwise} \end{cases} \tag{4.44}$$

This function returns zero when the x is a feasible solution of the problem. Now, we can define a new function B given by Equation (4.45):

$$B(x) = f(x) + I_g(x) \tag{4.45}$$

This function is similar to the Lagrangian. However, it is not continuous and cannot be optimized by simple derivation. Therefore, we define a continuous approximation for the indicator function as given in Equation (4.46). This function is called logarithmic barrier.

$$I_g(x) \approx \phi_\mu(x) = -\mu \ln(-g(x)) \tag{4.46}$$

Figure 4.5 shows the indicator function and the corresponding logarithmic barrier. The barrier function approximates the indicator function as $\mu \to 0$. The idea is to solve the problem using the approximated function using a

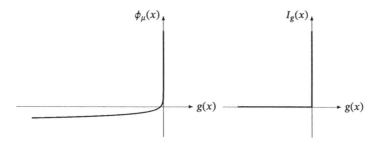

Figure 4.5 Indication function and logarithmic barrier.

continuous method (for example, Newton's method) and decrease the value of μ iterative until achieving convergence.

Although this is oversimplification of the algorithm implemented in practice, it is useful to understand the concept. There are several variants to the interior point algorithm. Interested readers can refer to [17] for more details.

4.8 Mixed-integer programming problems

A mixed-integer programming (MIP) problems is an optimization problem with both continuous and discrete variables. Discrete spaces are non-convex, and hence, all MIP problems are non-convex. More notably, MIPs are NP-hard which means they are among the most challenging problems in terms of theoretical complexity. However, some mixed-integer programming problems can be solved in Python using CvxPy with a suitable solver. The most common approach for solving this type of problem is the Branch & Bound (B&B) algorithm, which is based on the idea of dividing the problem until a binary solution is found.

Let us see the basic philosophy of the algorithm by considering the following binary optimization problem:

$$\min f(x)$$
$$x_i \in \mathbb{B} \tag{4.47}$$

where $f : \mathbb{R}^n \rightarrow \mathbb{R}$ is convex and $\mathbb{B} = \{0, 1\}$. First, we solve the following relaxed problem:

$$\min f(x)$$
$$0 \le x_i \le 1 \tag{4.48}$$
$$x_i \in \mathbb{R}$$

we would be lucky if this solution turns out to be binary. Most probably, the solution would be real values such as $x_i = 0.8$ or $x_i = 0.3$. This solution is obviously not feasible from the point of view of the binary problem. However, it is a lower bound f^{lower} of the problem. A binary upper bound is also required and marked as f^{upper}.

The B&B algorithm departs from f^{lower} and evaluates different problem instances using a branching rule. For example, we evaluate the solution with $x_i = 0$ and the solution with $x_i = 1$. If one of these branching problems results to be binary and higher than f^{upper}, then the solution is discarded as well as the branching stages below this instance. If the solution is lower than f^{upper}, then we have a new f^{upper} and continue the algorithm. The main drawback of this algorithm is the high computational load associated with evaluating each node, as the tree is built. Therefore, we require efficient branching rules to reduce the number of nodes that are evaluated. Most of the commercial solvers have additional techniques to accelerate the process. The following example shows how the algorithm works in practice.

Example 4.15. Consider the following optimization problem:

$$\min f(x) = x^{\mathsf{T}} H x$$
$$\sum x_i = 3 \tag{4.49}$$
$$x_i \in \mathbb{B}$$

with $H = \text{diag}(0.41, 0.51, 0.32, 0.20, 0.31, 0.21)$ and $i \in \{0, 1, \dots, 6\}$. First, we relax the binary constraint obtaining the following convex model:

$$A = \left\{ \begin{array}{c} \min x^{\mathsf{T}} H x \\ \sum x_i = 3 \\ 0 \le x_i \le 1 \end{array} \right\} \tag{4.50}$$

If the solution of this problem is binary, then our problem is solved. However, this is not the case, so A is just a lower bound. Then we generate new instances of the problem with $x_0 = 0$ and $x_1 = 1$, namely

$$B = \left\{ \begin{array}{c} \min x^{\mathsf{T}} H x \\ \sum x_i = 3 \\ 0 \le x_i \le 1 \\ x_0 = 0 \end{array} \right\} \tag{4.51}$$

and

$$C = \left\{ \begin{array}{c} \min x^{\mathsf{T}} H x \\ \sum x_i = 3 \\ 0 \le x_i \le 1 \\ x_0 = 1 \end{array} \right\} \tag{4.52}$$

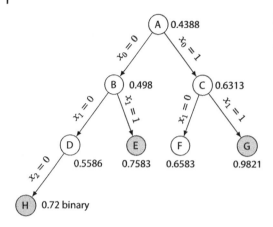

Figure 4.6 Tree generated by the branch and bound algorithm.

The process continues generating a tree as depicted in Figure 4.6. Eventually, the algorithm finds a binary solution; in this case, in the node H with $f = 0.72$ and $x = (0, 0, 0, 1, 1, 1)^\top$. This binary solution is an upper bound that blocks any solution resulting from nodes E and G. The branching process must be executed from nodes D and F until a better binary solution is found or until the nodes become higher than the upper bound. If a better binary solution is found, this is a set as the new upper bound, and the process continues until the tree is completed.

Table 4.1 shows in detail the first nodes generated by the algorithm for the states depicted in Figure 4.6. The enumeration tree generated by the branch and bound method may be large, and hence, the algorithm is not as efficient as the algorithms for continuous optimization. However, this is the primary tool for most of the integer and binary problems in practice.

Example 4.16. There are efficient solvers for binary problems available in CvxPy, so we do not require generating the enumeration tree by hand. The code for the previous example is presented below:

```
import numpy as np
import cvxpy as cvx
x = cvx.Variable(6, boolean=True)
H = np.diag([0.41,0.51,0.32,0.20,0.31,0.21])
f = cvx.Minimize(cvx.quad_form(x,H))
res = [ 0<= x, x<=1, cvx.sum(x)==3]
BinaryProblem = cvx.Problem(f,res)
BinaryProblem.solve()
print('Optimal value',np.round(f.value,4),np.round(x.value,4))
```

Table 4.1 Details of the node generated by the branch and bound problem.

Node	$f(x)$	x_0	x_1	x_2	x_3	x_4	x_5
A	0.4388	0.3567	0.2868	0.4570	0.7313	0.4718	0.6964
B	0.4980	0	0.3255	0.5187	0.8299	0.5354	0.7904
C	0.6313	1	0.2170	0.3458	0.5533	0.3570	0.5269
D	0.5586	0	0	0.5818	0.9309	0.6006	0.8866
E	0.7583	0	1	0.3879	0.6206	0.4004	0.5911
F	0.6583	1	0	0.3879	0.6206	0.4004	0.5911
G	0.9821	1	1	0.1939	0.3103	0.2002	0.2955
H	0.7200	0	0	0	1	1	1

The only difference concerning the continuous problem is that variable x is defined as binary with the parameter `boolean=True`.

Although it is easy to generate binary problems, we must keep in mind that each binary variable implies an increase in the enumeration tree's size, which makes the model more complex and the algorithm slower. As a basic rule, we should reduce, if possible, the number of binary variables in our models.

4.9 Transforming MINLP into MILP

Mixed-integer non-linear programming problems (MINLP) are among the most complicated mathematical optimization problems in theory and practice. Therefore, it is convenient to transform these models into mixed-integer linear programming problems (MILP). Below, we present some examples of heuristic transformations.

Example 4.17. Let us consider a set of constraints as presented below:

$$y = ux$$
$$x_{\text{low}} \leq x \leq x_{\text{up}} \qquad (4.53)$$
$$u \in \mathbb{B}, x \in \mathbb{R}, y \in \mathbb{R}$$

The first constraint is both non-linear, non-convex, and mixed-integer. Therefore, it is convenient to transform it into a set of mixed-integer affine constraints as follows:

$$ux_{\text{low}} \le y \le ux_{\text{up}}$$

$$x - (1-u)(x_{\text{up}} - x_{\text{low}}) \le y \le x + (1-u)(x_{\text{up}} - x_{\text{low}})$$

$$x_{\text{low}} \le x \le x_{\text{up}} \tag{4.54}$$

$$u \in \mathbb{B}, x \in \mathbb{R}, y \in \mathbb{R}$$

Notice that if $u = 0$ then the first constraint is reduced to $y = 0$, whereas if $u = 1$ the two fist constraints result in $x_{\text{low}} \le y \le x_{\text{up}}$ and $y = x$, respectively. These conditions are equivalent to $y = ux$ with $u \in \mathbb{B}$.

Example 4.18. Mixed-integer quadratic programming problems as the one given in Equation (4.49) can be easily transformed into a MILP problem. First, we write the quadratic in polynomial form as follows:

$$f(x) = \sum_k \sum_m h_{km} x_k x_m \tag{4.55}$$

Then, we notice that $x^2 = x$ for $x \in \mathbb{B}$. Therefore, the terms in the diagonal of the quadratic form can be replaced by linear equations as presented below:

$$f(x) = \sum_k h_{kk} x_k + \sum_k \sum_{m \ne k} h_{km} x_k x_m \tag{4.56}$$

Next, the bi-linear terms $x_k x_m$ are replaced by a new binary variable y_{km}, that is to say:

$$f(x) = \sum_k h_k k x_k + \sum_k \sum_{m \ne k} h_{km} y_{km} \tag{4.57}$$

Finally, we add the following auxiliary constraints:

$$x_k + x_m - 1 \le y_{km}$$

$$y_{km} \le x_k \tag{4.58}$$

$$y_{km} \le x_m$$

Notice that, under these constraints, $y_{km} = 1$ if $x_k = 1$ and $x_m = 1$, otherwise $y_{km} = 0$. The model is now an MIP problem.

4.10 Further readings

In this chapter, we learned how to solve linear and quadratic problems in Python using CvxPy. The reader is invited to see the module's manual in [23] and reproduce the examples presented there. Most of the solvers called by Python are modifications of the gradient, the Newton's, and/or the interior point method. These methods have well-defined conditions that guarantee convergence if the problem is convex. The theory behind these algorithms is beyond this book's objectives. It can be found in [11] and [24].

A reader interested in delving into the theory can continue with Chapter 5 where a family of convex optimization problems, known as conic optimization, is studied. A reader interested in applications for power systems operation can go directly to Chapter 7, which presents the economic dispatch of thermal units.

There is extensive literature about mixed-integer problems that go beyond the practical objectives of this book. Most of the solvers for mixed-integer optimization are based on the branch and bound method, although other algorithms such as the cutting plane and greedy algorithms are also used. A good presentation about these methods, both in theory and practice, is available in [25].

4.11 Exercises

1. Solve the following linear programming problem:

$$\min c^T x$$

$$\sum_i x_i = 1 \tag{4.59}$$

$$x_i \geq 0$$

where $c, x \in \mathbb{R}^n$ and $c_i = i + 1$. Solve the problem for $n = 2$, $n = 10$ and 100.

2. Solve the transportation problem with six sources and eight demands described in Table 4.2

3. Solve the following problem in Python using the module CvxPy.

$$\min 3x^2 + 2y^2 + 5z^2$$

$$x + y + z = 1 \tag{4.60}$$

$$x, y, z \geq 0$$

4. Solve the following problem similar to Example 4.13 for $n = 4$ and $n = 5$.

$$\min \sum_{i=0}^{n-1} x_i$$

$$x \in \mathcal{B}_a \cap \mathcal{B}_b \tag{4.61}$$

where $a = 1_n$ (i.e., a vector with all entries equal to 1) and $b = 0_n$ (i.e., a vector of zeros).

Table 4.2 Parameters for a transportation problem with six sources and eight demands.

c_{ij}	0	1	2	3	4	5	s_i
0	43	90	10	58	95	60	175
1	49	41	65	75	25	17	62
2	33	41	26	64	72	29	118
3	16	49	84	26	36	91	118
4	8	95	82	66	2	17	58
5	90	92	28	32	55	66	175
6	95	90	71	87	69	72	173
7	66	87	29	40	37	52	122
d_i	212	144	92	168	201	184	total $= 1001$

5. Solve the following optimization problem using Python and the module CvxPy

$$\min x^2 + y^2$$
$$(x - 1)^2 + (y - 1)^2 \leq 1 \tag{4.62}$$
$$(x - 1)^2 + (y + 1)^2 \leq 1$$

6. Solve the following quadratic optimization problem:

$$\min \frac{1}{2}(x - \mathbf{1}_n)^{\mathsf{T}} H(x - \mathbf{1}_n)$$
$$\sum x_i = 1 \tag{4.63}$$

where $\mathbf{1}_n$ is a vector with all entries equal to 1 and H is a symmetric matrix of size $n \times n$ constructed in the following way: $h_{km} = (m + k)/2$ if $k \neq m$ and $h_{km} = n^2 + n$ if $k = m$. Use $n = 2$, $n = 10$, and $n = 100$.

7. Solve the following optimization model using the basic interior point method described in Section 4.7.

$$\min x^2$$
$$x \geq 5 \tag{4.64}$$

8. A matrix A is diagonal dominant if its entries are such that

$$a_{kk} \geq \sum_{m \neq k} |a_{km}| \tag{4.65}$$

Show that every dominant diagonal matrix is positive semidefinite, but the opposite is not true.

9. Define a function in Python that generates a random positive definite matrix of size $n \times n$. Use this function to generate random matrices A and B; evaluate numerically each of the properties given in Section 4.4.

10. Finish Example 4.15 and compare the solution with Example 4.16.

5

Conic optimization

Learning outcomes

By the end of this chapter, the student will be able to:

- Identify the main features related to semidefinite and second-order cone optimization.
- Transform optimization problems into standard SDP or SOC models.
- Solve SDP and SOC problems using Python.

5.1 Convex cones

A cone is a set $\mathcal{C} \in \mathbb{R}^n$ such that if $x \in \mathcal{C}$ then $\alpha x \in \mathcal{C}$. A convex cone is set that is simultaneously a cone and a convex set as depicted in Figure 5.1. A conic optimization problem minimizes a convex function over the intersection of an affine subspace and a convex cone. Two particular types of convex cones are relevant in power systems operation: the cone generated by semidefinite matrices and the second-order cone. Linear, quadratic, and quadratically constrained problems can be considered as particular cases of these cones. In addition, there are several solvers available in CvxPy that efficiently solve conic optimization problems. The following sections studies theoretical and practical aspects of conic optimization in Python.

5.2 Second-order cone optimization

A second-order cone or SOC is a set in $\mathbb{R}^{n+1}$ given by the following expression:

$$\mathcal{C}_{\text{SOC}} = \left\{ (x, z) \in \mathbb{R}^{n+1} \ : \ \|x\| \leq z \right\} \tag{5.1}$$

Mathematical Programming for Power Systems Operation: From Theory to Applications in Python. First Edition. Alejandro Garcés.
© 2022 by The Institute of Electrical and Electronics Engineers, Inc. Published 2022 by John Wiley & Sons, Inc.

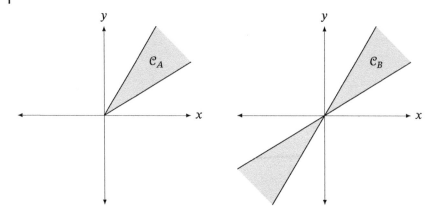

Figure 5.1 Example of a convex cone $\mathcal{C}_A$ and a non-convex cone $\mathcal{C}_B$

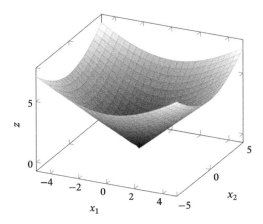

Figure 5.2 Representation of the second order cone $\Omega = \{\|x\| \leq z\}$ with $x \in \mathbb{R}^2$ and $z \in \mathbb{R}$.

where x is a vector in $\mathbb{R}^n$; z is a real variable; and $\|\cdot\|$ is the Euclidean norm, given by Equation (5.2).

$$\|x\| = \sqrt{x_0^2 + x_1^2 + x_2^2 + \cdots + x_{n-1}^2} \tag{5.2}$$

Figure 5.2 depicts the boundary of an SOC with $x \in \mathbb{R}^2$.

Let us see why Equation (5.1) defines a cone. Consider a point $(x, z) \in \mathcal{C}_{\text{SOC}}$ and a positive scalar α. Now, consider a point scaled by that α. This new point holds the same inequality as given below:

$$\|\alpha x\| \leq \alpha z \tag{5.3}$$

In other words, the point $(\alpha x, \alpha z) \in \mathcal{C}_{SOC}$. Therefore, $\mathcal{C}_{SOC}$ is a cone. It is straightforward to demonstrate that this cone is also convex using the properties of the norm (Section 2.2) and the Jensen's inequality.

On the other hand, both sides of the inequality in Equation (5.1) can be composed by affine spaces obtaining a general representation of an SOC constraint, as given below:

$$\|Ax + b\| \leq c^T x + d \tag{5.4}$$

where A is a matrix, b and c are vectors, and d is a scalar. Thus, a general representation of an SOC optimization problem is the following:

$$\min h^T x$$
$$\|Ax + b\| \leq c^T x + d \tag{5.5}$$

However, many practical problems are a combination of different types of constraints. Therefore, a practical problem might not be an SOC optimization problem but a convex problem with second-order cone constraints.

Among convex optimization problems, SOC optimization is particularly appealing in practice for three main reasons: first, there are several available algorithms and solvers for SOC problems. These algorithms are fast and efficient in practice. Second, several optimization problems may be represented as SOC problems. For example, a linear programming problem is a particular case of an SOC problem with $A = 0$, and a quadratic-convex restriction is also presentable as an SOC constraint. Third, it is possible to transform some non-convex problems into equivalent SOC problems. The examples below show some of these equivalents.

Example 5.1. A convex-quadratic constraint can be represented as an SOC. Let us consider the following inequality that is convex if $M > 0$:

$$x^T M x - n^T x - s \leq 0 \tag{5.6}$$

Since $M > 0$ (see Section 4.3) then it has Cholesky factorization given by $M = (M^{\frac{1}{2}})^T M^{\frac{1}{2}}$.

Let us define two new variables u and z as follows:

$$u = M^{\frac{1}{2}} x \tag{5.7}$$
$$z = n^T x + s \tag{5.8}$$

then, Equation (5.6) is equivalent to the following constraint:

$$u^T u \leq z \tag{5.9}$$

This inequality can be transformed into an SOC by adding $z^2/4 - z/2 + 1/4$ in both sides of the inequality, as given below:

$$u^\top u + \frac{z^2}{4} - \frac{z}{2} + \frac{1}{4} \leq z + \frac{z^2}{4} - \frac{z}{2} + \frac{1}{4} \tag{5.10}$$

$$u^\top u + \left(\frac{z-1}{2}\right)^2 \leq \left(\frac{1+z}{2}\right)^2 \tag{5.11}$$

$$\left\|\begin{pmatrix} u \\ \frac{z-1}{2} \end{pmatrix}\right\| \leq \frac{1+z}{2} \tag{5.12}$$

Returning to the original variables, we have the following expression:

$$\left\|\begin{pmatrix} M^{\frac{1}{2}}x \\ \frac{n^\top x + s - 1}{2} \end{pmatrix}\right\| \leq \frac{1 + n^\top x + s}{2} \tag{5.13}$$

By using this method, any convex-quadratic model can be transformed into an SOC problem.

Example 5.2. Consider the following set of inequality constraints:

$$xy \geq w^\top w$$
$$x \geq 0 \tag{5.14}$$
$$y \geq 0$$

where w is a vector $\in \mathbb{R}^n$ and x, y are variables in $\mathbb{R}$. These inequalities define a hyperbolic set that, at fist glance looks like a non-convex set. However, it is in fact convex and can be transformed into an SOC; for this, we start from the following representation of an hyperbolic paraboloid:

$$xy = \frac{1}{4}(x+y)^2 - \frac{1}{4}(x-y)^2 \tag{5.15}$$

then, we have the following:

$$xy \geq w^\top w \tag{5.16}$$

$$\frac{1}{4}(x+y)^2 - \frac{1}{4}(x-y)^2 \geq w^\top w \tag{5.17}$$

$$(x+y)^2 \geq (x-y)^2 + 4w^\top w \tag{5.18}$$

the last inequality can be easily transformed into an SOC as follows:

$$x + y \geq \left\|\begin{pmatrix} 2w \\ x - y \end{pmatrix}\right\| \tag{5.19}$$

This example shows that SOC is a very general way to represent many non-linear problems.

Example 5.3. Consider the following set:

$$z = xy$$
$$0 \leq x \leq 1 \tag{5.20}$$
$$0 \leq y \leq 1$$

A constraint of the form $z = xy$ is not convex. Therefore, it is common to use a linearization to include this type of constraints into an optimization problem, namely:

$$z \geq 0$$
$$x \leq z$$
$$y \leq z \tag{5.21}$$
$$z \geq x + y - 1$$

This linearization works in most of the cases. However, for a constraint of the form $z^2 = xy$, it is more precise to use a SOC approximation. We transform the equality into an inequality, and use Equation (5.19) as follows:

$$\left\| \begin{pmatrix} 2z \\ x - y \end{pmatrix} \right\| \leq x + y$$
$$0 \leq x \leq 1 \tag{5.22}$$
$$0 \leq y \leq 1$$
$$z \geq 0$$

The SOC approximation maintains the non-linear nature of the problem, making it convex. Notice the point $(1/2, 1/2, 1/4)$ is feasible in both the original problem and the SOC approximation. However, it is infeasible in linearization.

Example 5.4. The function $f(x) = -\ln(x)$ is convex and hence the following set is also convex:

$$-\ln(x + 1) \leq z$$
$$-\frac{1}{2} \leq x \leq \frac{1}{2} \tag{5.23}$$

However, it is possible to obtain an SOC approximation of this set by considering a quadratic expansion of the logarithmic function:

$$\ln(x+1) \approx x - \frac{x^2}{2} \tag{5.24}$$

This approximation can be included into Equation (5.23) obtaining the following SOC constraint:

$$\left\| \begin{array}{c} x \\ x+z-1/2 \end{array} \right\| \leq \frac{1+2(x+z)}{2}$$

$$-\frac{1}{2} \leq x \leq \frac{1}{2} \tag{5.25}$$

The student is invited to plot the set given by Equation (5.23) and the set given by Equation (5.25) and compare these results.

Example 5.5. A SOC problem can be easily solved in Python. Consider the following optimization model:

$$\min z$$

$$\|Ax+b\| \leq z \tag{5.26}$$

where $x \in \mathbb{R}^n$, $A \in \mathbb{R}^{m \times n}$, and $b \in \mathbb{R}^m$. The following code solves a random instance of this problem, for $n = 10$ and $m = 6$:

```
import numpy as np
import cvxpy as cvx
n = 10
m = 6
A = np.random.rand(m,n)
b = np.random.rand(m)
x = cvx.Variable(n)
z = cvx.Variable()
obj = cvx.Minimize(z)
res = [cvx.SOC(z,A@x+b)]
prob = cvx.Problem(obj, res)
prob.solve()
```

The key function in this example, is the SOC constraint presented in the penultimate line. As always, the reader is invited to experiment with this code.

5.2.1 Duality in SOC problems

Duality theory is easily applied to second-order cone optimization problems. Consider the SOC model given by Equation (5.5) which is equivalent to the model presented below:

$$\min h^\mathsf{T} x$$

$$\|u\| \le c^\mathsf{T} x + d \tag{5.27}$$

$$u = Ax + b$$

We define a Lagrangian function as follows:

$$\mathcal{L}(x, u, y, z) = h^\mathsf{T} x + y(\|u\| - c^\mathsf{T} x - d) + z^\mathsf{T}(u - Ax - b) \tag{5.28}$$

with $y \ge 0$. Now, we take the infimum in x and u, in order to obtain the dual function. Fortunately, the problem is separable as presented below:

$$\inf_{x,u} \mathcal{L} = \inf_{x,u} \left(h^\mathsf{T} x + y(\|u\| - c^\mathsf{T} x - d) + z^\mathsf{T}(u - Ax - b) \right) \tag{5.29}$$

$$= \inf_{x,u} \left((-yd - b^\mathsf{T} z) + (h^\mathsf{T} - yc^\mathsf{T} - z^\mathsf{T} A)x + (y \|u\| + z^\mathsf{T} u) \right) \tag{5.30}$$

$$= -yd - b^\mathsf{T} z + \inf_x (h^\mathsf{T} - yc^\mathsf{T} - z^\mathsf{T} A)x + \inf_u (y \|u\| + z^\mathsf{T} u) \tag{5.31}$$

Thus $\inf_x \mathcal{L}(x, u, y, z)$ implies that:

$$h - yc - A^\mathsf{T} z = 0 \tag{5.32}$$

and $\inf_u \mathcal{L}(x, u, y, z)$ is obtained from:

$$\inf_u y \|u\| + z^\mathsf{T} u \tag{5.33}$$

If we are using the Euclidean norm, then the Cauchy inequality is valid:

$$|z^\mathsf{T} u| \le \|z\| \|u\| \tag{5.34}$$

and consequently

$$- \|z\| \|u\| \le z^\mathsf{T} u \le \|z\| \|u\| \tag{5.35}$$

$$- \|z\| \|u\| + y \|u\| \le y \|u\| + z^\mathsf{T} u \le \|z\| \|u\| + y \|u\| \tag{5.36}$$

$$(y - \|z\|) \|u\| \le y \|u\| + z^\mathsf{T} u \le (y + \|z\|) \|u\| \tag{5.37}$$

The infimum in both the right and the left-hand side of this inequality is zero as long as $y - \|z\| \ge 0$. Combining all these results, we have the following dual problem:

$$\min yd + b^\mathsf{T} z$$
$$yc + A^\mathsf{T} z = h \tag{5.38}$$
$$\|z\| \leq y$$

In conclusion, the dual of an SOC problem is another SOC problem. Since an SOC problem is convex, then we can conclude the dual is equal to the primal as long as it fulfills Slater's conditions.

5.3 Semidefinite programming

Another important cone for mathematical optimization is the cone generated by positive semidefinite matrices. Here we are interested in solving problems with the following structure:

$$\min \ \operatorname{tr}(CX)$$
$$AX = B \tag{5.39}$$
$$X \succeq 0$$

where A, B, C, X are matrices. This problem looks very different from the convex problems we have studied so far; decision variables are now matrices $X \in \mathbb{R}^{n \times n}$ and the objective function is the trace of a matrix product. More importantly, there is a constraint of form $X \succeq 0$ that indicates the matrix is positive-semidefinite[1].

Before studying these problems, let us review some basic concepts from matrix algebra.

5.3.1 Trace, determinant, and the Shur complement

The trace is an operator that takes a square matrix A and returns a scalar equal to the sum of the entries in the diagonal, as follows:

$$\operatorname{tr}(A) = a_{11} + a_{22} + \dots a_{nn} \tag{5.40}$$

This operator have some useful properties, namely

- $\operatorname{tr}(A + B) = \operatorname{tr}(A) + \operatorname{tr}(B)$
- $\operatorname{tr}(\alpha A) = \alpha \operatorname{tr}(A)$
- $\operatorname{tr}(A) = \operatorname{tr}(A^\mathsf{T})$
- $\operatorname{tr}(A^\mathsf{T} B) = \operatorname{tr}(AB^\mathsf{T})$

---

1 Notice the symbol is different from an inequality.

- $\mathrm{tr}(AB) = \mathrm{tr}(BA)$ but in general, $\mathrm{tr}(AB) \neq \mathrm{tr}(A)\,\mathrm{tr}(B)$
- si $A \succ 0$ y $B \succ 0$ then $\mathrm{tr}(AB) \geq 0$
- $x^\mathsf{T} H x = \mathrm{tr}(H x x^\mathsf{T})$
- $\mathrm{tr}(H) = \sum \lambda_i$ where λ_i are the eigenvalues of H

Example 5.6. Let us experiment in Python by taking random matrices A, B and checking the aforementioned properties:

```
import numpy as np
n = 10
A = np.random.rand(n,n)
B = np.random.rand(n,n)
print("sum",np.trace(A+B), np.trace(A)+np.trace(B))
print("prd",np.trace(A@B), np.trace(A)*np.trace(B))
```

The determinant is another operator that takes a square matrix and returns a scalar. It has the following properties:

- A matrix A has a unique inverse if $\det(A) \neq 0$
- $\det(AB) = \det(A)\det(B)$ but in general $\det(A + B) \neq \det(A) + \det(B)$
- $\det(A^\mathsf{T}) = \det(A)$
- If a matrix is triangular, then the determinant is the product of the entries in the main diagonal.
- $\det(A^{-1}) = 1/\det(A)$
- $\det(A) = \det(P^{-1}AP)$
- $\det(A) = \prod \lambda_i$ where $\lambda = \mathrm{eig}(A)$

These are useful properties of the determinant, which is a complex and interesting operator[2]. The following example shows a geometric interpretation of the determinant.

Example 5.7. As we have seen, a quadratic form $x^\mathsf{T} H x$ with $H \succ 0$ may have different interpretations. For instance, it can generate an ellipsoid $\mathcal{E}$ defined as follows:

$$\mathcal{E} = \{x \in \mathbb{R}^n : x^\mathsf{T} H x \leq 1\} \tag{5.41}$$

Since H is positive definite, we can make a Cholesky factorization, namely:

$$H = (H^{\frac{1}{2}})^\mathsf{T} H^{\frac{1}{2}} \tag{5.42}$$

---

2 We suppose the student is familiar with the ways as the determinant is calculated. A student interested in a formal definition can refer to [21] pg 632.

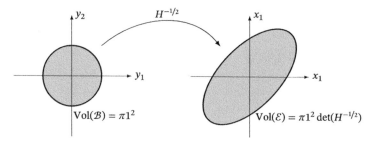

Figure 5.3 Area of an ellipsoid seen as a linear transformation of a unit ball

that allows to define the following linear transformation $y = H^{\frac{1}{2}}x$ that in turns, transforms the ellipsoid as follows $\mathcal{B} = \{y \in \mathbb{R}^n : y^\mathsf{T}y \le 1\}$. This is a unitary ball of which we know its hypervolume[3]. The determinant allows to calculate the volume of the ellipsoid as shown in Figure 5.3.

To end this review of matrix algebra, let us define the Shur complement. Consider the following block matrix

$$M = \begin{pmatrix} A & B \\ B^\mathsf{T} & C \end{pmatrix} \tag{5.43}$$

then, we define the Shur complement of M with respect to C as follows:

$$A - BC^{-1}B^\mathsf{T} \tag{5.44}$$

If $M \succeq 0$ and $A \succeq 0$, then its Shur complement is also positive semidefinite, and vice versa.

We can also define the Shur complement respect to A as given below:

$$C - B^\mathsf{T}A^{-1}B \tag{5.45}$$

The Shur complement is useful because we can identify when a matrix is positive semidefinite by evaluating the condition in some sub-matrices.

---

3 The hypervolume is the generalization of measurements such as length, area, and volume for $\mathbb{R}^n$. The hypervolume of a unitary ball in $\mathbb{R}^2$ is the area, i.e., $\pi 1^2$.

5.3.2   Cone of semidefinite matrices

The set of semidefinite matrices defines a convex cone. To prove it, consider the following set:

$$\Omega = \{X \in \mathbb{R}^{n \times n} \; : \; X \succeq 0\} \tag{5.46}$$

This set defines a cone since if $X \in \Omega$ then αX is also in Ω (αX is positive semidefinite if $X \succeq 0$ and $\alpha \geq 0$). Now consider two matrices $X, Y \in \Omega$ and two scalars α, β such that $\alpha + \beta = 1$ and $\alpha, \beta \geq 0$. Since an intermediate matrix $Z = \alpha X + \beta Y$ also belongs to Ω, we conclude the set is convex.

A semidefinite programming problem or SDP is any problem that includes semidefinite constraints as in Equation (5.39). Far from being a meaningless or obscure mathematical theory, SDP is a practical tool for different optimization problems, as shown in the following examples.

Example 5.8. A SOC is a particular case of an SDP. Consider the following SOC:

$$\|u\| \leq t \tag{5.47}$$

which is equivalent to

$$u^{\top} u \leq t^2 \tag{5.48}$$

$$t^2 - u^{\top} I u \geq 0 \tag{5.49}$$

where I is the identity matrix. This constraint can be transformed into a semidefinite constraint using the Shur complement as follows:

$$\begin{pmatrix} tI & u \\ u^{\top} & t \end{pmatrix} \succeq 0 \tag{5.50}$$

As a direct consequence of this, there is a hierarchy among SOC, SDP, QP, and LP problems as given in Figure 5.4, since linear programming (LP), quadratic programming (QP), and quadratically constrained quadratic programming (QCQP) are particular cases of SOC.

🐍

Example 5.9. Semidefinite programming is of practical interest because it is possible to solve SDP problems using efficient algorithms. However, some problems can be represented either as SDP or SOC problems. Below, we present the solution in Python of a random instance of the following optimization problem:

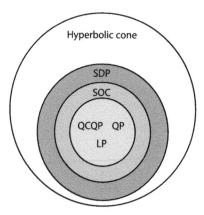

Figure 5.4 Venn diagram that represents the relation among different connic problems

$$\min t$$

$$c^{\mathsf{T}} u \geq 1 \tag{5.51}$$

$$\|u\| \leq t$$

```
import numpy as np
import cvxpy as cvx
n = 10
c = np.random.rand(n)
u = cvx.Variable(n)
t = cvx.Variable()
obj = cvx.Minimize(t)
res = [c@u >= 1, cvx.SOC(t,u)]
ModelSOC = cvx.Problem(obj,res)
ModelSOC.solve()
print("SOC:",obj.value)
print("Time:",ModelSOC.solver_stats.solve_time)

I = np.eye(n)
X = cvx.Variable((n+1,n+1),symmetric=True)
obj = cvx.Minimize(t)
res = [c@u >= 1, X >>0]
res += [X[n,n] == t]
for k in range(n):
  res += [ X[k,n] == u[k]]
  res += [ X[n,k] == u[k]]
  for m in range(n):
    res += [X[k,m] == t*I[k,m]]
ModelSDP = cvx.Problem(obj,res)
```

```
ModelSDP.solve()
print("SDP:",obj.value)
print("Time:",ModelSDP.solver_stats.solve_time)
```

A semidefinite constraint is generated by a new variable X and the symbol $>>$ that indicates positive semidefinite[4]. It is interesting to see the difference in time calculation for SDP and SOC problems. Usually, SOC is solved more efficiently than SDP. The student is invited to experiment with different sizes of the problem, changing the values of n.

Example 5.10. Semidefinite programming is a highly flexible tool for modeling; as we have seen, SOC, QP, and LP are particular SDP cases. Many other optimization problems can be represented as SDP models. Table 5.1 shows a list of optimization constraints that can be transformed into an SDP constraint.

Table 5.1 Constraints that can be transformed to SDP.

Constraint	Semidefinite constraint
$\|u\| \leq t$	$\begin{pmatrix} tI & u \\ u^\mathsf{T} & t \end{pmatrix} \succeq 0$
$x^2 + y^2 \leq 1$	$\begin{pmatrix} 1+x & y \\ y & 1-x \end{pmatrix} \succeq 0$
$\frac{(c^\mathsf{T}x)^2}{d^\mathsf{T}x}$	$\begin{pmatrix} d^\mathsf{T}x & c^\mathsf{T}x \\ c^\mathsf{T}x & t \end{pmatrix} \succeq 0$

5.3.3 Duality in SDP

Duality theory can be easily extended to SDP problems, for this, consider the following problem:

---

4 It is essential to remember that $>>$ is different from $>=$. The first symbol indicates the matrix is positive semidefinite, whereas the second means that all the matrix entries are positive.

$$\min \ \operatorname{tr}(CX)$$

$$\operatorname{tr}(A_i X) = b_i \ \forall i \tag{5.52}$$

$$X \succeq 0$$

where affine constraints are represented using a trace function. Let us define a Lagrangian funtion as follows:

$$\mathcal{L}(X, z, Y) = \operatorname{tr}(CX) + \sum_i z_i(b_i - \operatorname{tr}(A_i X)) - \operatorname{tr}(YX) \tag{5.53}$$

where Y is a square-symmetric and positive semidefinite matrix $(Y = Y^\mathsf{T} \succeq 0)$ and z is a vector. We can define the dual function just as in the case of general convex optimization problems, namely:

$$W(z, Y) = \inf_x \mathcal{L}(x, z, Y) \tag{5.54}$$

The following conditions are required to guarantee the existence of this infimum[5]:

$$\operatorname{tr}(CX) - \sum_i z_i \operatorname{tr}(A_i X) - \operatorname{tr}(YX) = 0 \tag{5.55}$$

$$\operatorname{tr}\left(CX - \sum_i z_i A_i X - YX\right) = 0 \tag{5.56}$$

$$\therefore \ \sum_i z_i A_i + Y = C \tag{5.57}$$

therefore, the dual problem takes the following form

$$\max \ b^\mathsf{T} z$$

$$Y + \sum_i z_i A_i = C \tag{5.58}$$

$$Y \succeq 0$$

Just as in any optimization problem, the dual problem is such that dual $\leq$ primal and the Slater conditions.

5.4 Semidefinite approximations

In Example 5.3, we showed how a hyperbolic constraint might be transformed into a second-order cone constraint. Similarly, we can convert some non-convex quadratic constraints into semidefinite thereof.

---

5 Here, we use the fact that the trace is distributive with respect to the sum.

Let us consider the following quadratic equality constraint:

$$x^T H x = 1 \tag{5.59}$$

where H is a square matrix. This constraint is evidently non-convex, even in the case of H semidefinite (recall equality constraints must be affine). In order to find a semidefinite approximation, we define a new matrix $X = xx^T$, namely:

$$X = xx^T = \begin{pmatrix} x_1 x_1 & x_1 x_2 & \cdots & x_1 x_n \\ x_2 x_1 & x_2 x_2 & \cdots & x_2 x_n \\ \vdots & \vdots & & \vdots \\ x_n x_1 & x_n x_2 & \cdots & x_n x_n \end{pmatrix} \tag{5.60}$$

We can express the quadratic form as function of this matrix as given below[6]:

$$x^T H x = \text{tr}(HX) \tag{5.61}$$

Evidently, X is positive semidefinite and $\text{rank}(X) = 1$, then the following constraint is equivalent to Equation (5.59):

$$\begin{aligned} \text{tr}(HX) &= 1 \\ X &\succeq 0 \\ \text{rank}(X) &= 1 \end{aligned} \tag{5.62}$$

This set is convex except for the rank constraint. Therefore, we can generate a convex approximation by relaxing this constraint. Notice the semidefinite condition is imposed in X and not in H. Thus, the approximation is very general for quadratic and hyperbolic problems.

Example 5.11. Consider the following quadratic optimization problem with quadratic constraints:

$$\begin{aligned} \min \ & x^T Q x \\ & x^T x = 1 \end{aligned} \tag{5.63}$$

with $x \in \mathbb{R}^2$ and $Q = Q^T \in \mathbb{R}^{2 \times 2}$ given below:

$$Q = \begin{pmatrix} 1 & 0.3 \\ 0.3 & -2 \end{pmatrix} \tag{5.64}$$

This problem is evidently non-convex; however, given its small size, it can be solved easily using Lagrange multipliers. To do so, we define the following

---

6 Recall properties given in Section 5.3.1

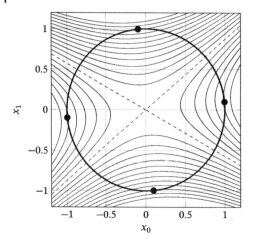

Figure 5.5 Graphical representation for the problem Equation (5.63): constraint (dark line), level curves of the objective function (light lines)

Lagrangian function:

$$\mathcal{L}(x, \lambda) = x^\top Q x + \lambda(1 - x^\top x) \tag{5.65}$$

with the following optimal conditions

$$Q x - \lambda x = 0 \tag{5.66}$$

$$x^\top x = 1 \tag{5.67}$$

Therefore, we can conclude that λ are the eigenvalues of Q and x are the unitary eigenvectors of Q. The optimal solution is just the minimum eigenvalue. Figure 5.5 shows the level curves of the objective function and the equality constraint in the space x_0, x_1. The problem has one minimum and one maximum, achieved at different points; that is to say, the solution is not unique.

This problem was easy to solve because it was defined in $\mathbb{R}^2$; the situation becomes more and more complex as n increases since we require to evaluate all possible eigenvalues and their corresponding eigenvectors.

Let us solve the problem using a convex approximation. We define a new matrix $X = x x^\top$ and the following optimization model which is completely equivalent to Equation (5.63):

$$
\begin{aligned}
\min \ & \operatorname{tr}(QX) \\
& \operatorname{tr}(X) = 1 \\
& X \succeq 0 \\
& \operatorname{rank}(X) = 1
\end{aligned}
\tag{5.68}
$$

Now, we relax the rank constraint obtaining a semidefinite optimization problem with the following coding in Python:

```python
import numpy as np
import cvxpy as cvx
Q = np.array([[1,0.3],[0.3,-2]])
X = cvx.Variable((2,2),symmetric=True)
fo = cvx.Minimize(cvx.trace(Q@X))
re = [X >> 0, cvx.trace(X)==1]
SDaprox = cvx.Problem(fo,re)
SDaprox.solve()
print('Aprox',fo.value)
print('Optimal',np.linalg.eigvals(Q))
```

We have transformed Model Equation (5.63) with two decision variables to Model Equation (5.68) with 4 decision variables. However, Model Equation (5.68) is convex (after relaxing the rank constraint) whereas Model Equation (5.63) is not. Finding a convex model improves the problem; in most applications, we prefer a large convex model instead of a medium-size non-convex problem.

Example 5.12. Binary problems may be solved using semidefinite approximations. Consider the following quadratic problem with binary constraints

$$\min\ x^T Q x$$
$$x_i \in \{-1, 1\} \tag{5.69}$$

In this case, the problem is binary although it takes values of $x_i = \pm 1$ instead of $0, 1$ as usual. This binary constraint is transformed into a quadratic equality constraint that is equivalent:

$$x_i^2 = 1 \tag{5.70}$$

Now we proceed as in the previous example obtaining a semidefinite model:

$$\min\ \text{tr}(QX)$$
$$\text{diag}(X) = 1 \tag{5.71}$$
$$X \succeq 0$$

This approximation may be efficient in some problems, for example, in the max-cut problem [26, 27].

5.5 Polynomial optimization

A polynomial optimization problem is a model that can be represented as follows:

$$\min f(x)$$
$$p(x) = 0 \tag{5.72}$$
$$q(x) \leq 0$$

Where f, p, and q are multivariable polynomials. These types of problems are, in principle, non-convex and highly complex. Polynomial optimization is very general since it encompasses linear, quadratic, and second-order optimization problems as well as many non-convex problems. In addition, binary constraints can be also transformed into polynomial constraints. For example, a binary variable that takes values $x = \pm 1$, may be represented as $x^2 = 1$. More importantly, polynomial optimization problems can be efficiently transformer into semidefinite optimization problems, as presented in this section.

A particular type of polynomials is those that can be represented as a sum-of-square (SOS), namely:

$$p(x) = \sum_i (q_i(x))^2 \tag{5.73}$$

Where q_i are multivariable polynomials. This type of problem can be transformed into semidefinite equivalents that are easier to solve. Below, we present series of examples that show how to transform SOS problems into SDP. Our exposition is practically oriented and based on the work presented by Parrillo et al. in [28].

The main idea, is that any polynomial SOS can be transformed as the equation presented below:

$$p(x) = (Qm(x))^\top (Qm(x)) \tag{5.74}$$

where $m(x)$ is a vector of monomials associated to p, and Q is a positively semidefinite matrix.

Example 5.13. Let us consider the following multivariate polynomial, namely:

$$p(x, y) = 2x^4 + 5x^2 + y^2 + 2yx^2 - 2xy \tag{5.75}$$

This polynomial is SOS since it can be represented as follows:

$$p(x,y) = \begin{pmatrix} x \\ y \\ x^2 \end{pmatrix}^T \begin{pmatrix} 5 & -1 & 0 \\ -1 & 1 & 1 \\ 0 & 1 & 2 \end{pmatrix} \begin{pmatrix} x \\ y \\ x^2 \end{pmatrix} \tag{5.76}$$

The vector of monomials is $m(x,y) = (x,y,x^2)^T$ and a matrix Q is defined as follows:

$$p(x,y) = \begin{pmatrix} x \\ y \\ x^2 \end{pmatrix}^T \begin{pmatrix} q_{00} & q_{01} & q_{02} \\ q_{10} & q_{11} & q_{12} \\ q_{20} & q_{21} & q_{22} \end{pmatrix} \begin{pmatrix} x \\ y \\ x^2 \end{pmatrix} \tag{5.77}$$

Therefore, the polynomial may be written as presented below:

$$p(x,y) = q_{00}x^2 + (q_{01}+q_{10})xy + (q_{02}+q_{20})x^3 + q_{11}y^2 + (q_{12}+q_{21})yx^2 + q_{22}x^4 \tag{5.78}$$

Matching term to term, a set of equality constraints is obtained as follows:

$$q_{00} = 5$$
$$q_{01} + q_{10} = -2$$
$$q_{02} + q_{20} = 0$$
$$q_{11} = 1 \tag{5.79}$$
$$q_{12} + q_{21} = 2$$
$$q_{22} = 2$$
$$Q \geq 0$$

This is a feasibility problem that can be solved in Python, as presented below:

```
import cvxpy as cvx
Q = cvx.Variable((3,3))
obj = cvx.Minimize(0)
res = [Q[0,0] == 5,
       Q[0,1]+Q[1,0] == -2,
       Q[0,2]+Q[2,0] == 0,
       Q[1,1] == 1,
       Q[1,2]+Q[2,1] == 2,
       Q[2,2] == 2,
```

```
        Q >> 0.01]
SOS = cvx.Problem(obj,res)
SOS.solve()
print(np.round(Q.value,3))
```

The result of this script is the following positive semidefinite matrix

$$Q = \begin{pmatrix} 5 & -1 & 0 \\ -1 & 1 & 1 \\ 0 & 1 & 2 \end{pmatrix} \tag{5.80}$$

The eigenvalues of this matrix are $\lambda = (5.25, 0.23, 2.52)$, so it is a positively semidefinite matrix. Therefore, $p(x, y)$ is a sum-of-square polynomial.

Example 5.14. Let us consider the polynomial given in Equation (5.81),

$$p(x) = x^4 - 30x^2 + 8x - 15 \tag{5.81}$$

A plot of this polynomial is depicted in Figure 5.6.

From the plot, we can conclude that the function is non-convex and have two local minimum at $x = -3.94$ and $x = 3.8$, with $p(-3.94) \approx -271.25$ and $p(3.8) \approx -209.28$, respectively. The global optimum is therefore ≈ -271.25.

Now, we formulate the problem as an approximated SOS optimization model, as presented below:

$$\min t$$
$$x^4 - 30x^2 + 8x - 15 - t \in SOS \tag{5.82}$$

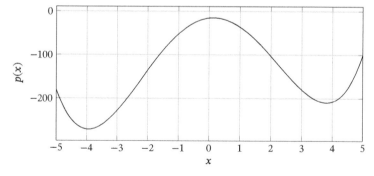

Figure 5.6 Plot of polynomial $p(x) = x^4 - 30x^2 + 8x - 15$

The SOS factorization of this polynomial is given below:

$$
\begin{pmatrix} 1 \\ x \\ x^2 \end{pmatrix}^{\mathsf{T}} \begin{pmatrix} q_{11} & q_{12} & q_{13} \\ q_{21} & q_{22} & q_{23} \\ q_{31} & q_{32} & q_{33} \end{pmatrix} \begin{pmatrix} 1 \\ x \\ x^2 \end{pmatrix} \tag{5.83}
$$

Matching term by term, we obtain the following equivalent semidefinite programming model:

$$
\begin{aligned}
\max \; & t \\
& Q = Q^{\mathsf{T}} \geq 0 \\
& q_{33} = 1 \\
& 2q_{23} = 0 \\
& q_{22} + 2q_{13} = -30 \\
& 2q_{12} = 8 \\
& q_{11} = -15 - t
\end{aligned} \tag{5.84}
$$

The optimal solution to this problem is obtained using CvxPy. The optimal value corresponds to -271.2461. That is the expected valued, according to Figure 5.6. Therefore, we have found the global optimum of the problem. The critical step in this problem was to transform a question from the multivariate polynomials' space to the positive semidefinite matrices' space and solve the resulting model.

5.6 Further readings

Second-order cone optimization and semidefinite programming are two of the most common type of conic optimization problems. Other applications and theoretical details can be studied in [29] and [30]. A useful review of linear algebra is essential to understand these concepts. See for example [16] and [21] where there is an excellent presentation of basic concepts such as determinant and quadratic forms. This chapter showed fundamental mathematical aspects. However, conic optimization has several power systems applications, especially in the optimal power flow problem. This aspect will be presented in Chapter 10.

Some aspects associated with polynomial optimization and, in particular, the sum-of-squares problems were also presented in this chapter. A complete study of this subject from the point of view of algebraic geometry is found in [28].

Importantly, SOS is not the unique type of polynomial optimization problem. However, SOS problems have the advantage of being representable as SDP and, therefore, they can be solved efficiently by means of algorithms of convex optimization, such as the interior point method. Another approach closely related to SOS and polynomial optimization is the Lasserre hierarchy. A good review of this subject can be found in [31] and [32].

5.7 Exercises

1. Make a function in Python that generates random square-semidefinite matrices of size $n \times n$.
2. Generate random instances of the following quadratically constrained quadratic program:

$$\min x^\mathsf{T} H x + r^\mathsf{T} x$$
$$x^\mathsf{T} M x + b^\mathsf{T} x + c \leq 0 \tag{5.85}$$

   Where H and M are positive semidefinite. Transform the problem into an SOC model and solve in Python.
3. Generate an example in Python for each of the cases given in Table 5.1.
4. Make a plot of the average execution time vs n for both the SOC and the SDP representation. Analyze the results.
5. A linear programming problem can be represented as an SDP problem. Generate a random instance of a linear programming problem and transform it in an SDP. Implement the code in Python and analyze the results.
6. A simple way to calculate an area, volume or hyper volume is by means of Monte Carlo integration. Suppose we are interested in calculating the shadow area in Figure 5.7, then we proceed as follows: first, we initialize a variable $s = 0$, next, we generate random values of x, y such that $-x_{\max} \leq x \leq x_{\max}$ and $-y_{\max} \leq y \leq y_{\max}$; if the point (x, y) belongs to the are a (such as the area A) then $s \leftarrow s + 1$; we repeat this procedure for n iterations (n requires to be a high value in order to obtain a good approximation); the area can be calculated as $(s/n)(4x_{\max}y_{\max})$. Use this procedure to calculate the area of the ellipsoid given in Example 5.7 for a random positive definite matrix H. Analyze also the case for calculating the volume of an ellipsoide in $\mathbb{R}^3$ and a hypervolume in $\mathbb{R}^4$.
7. The exponential function is convex, however, it can be approximated to an SOC constraint. Use a Taylor expansion around 0 and generate an SOC

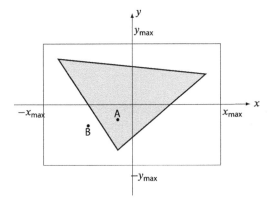

Figure 5.7 Example of a Monte Carlo integration. We add only the points inside the area that requires to be calculated

approximation of the following set:

$$\exp(x) \leq z$$
$$\frac{1}{2} \leq x \leq \frac{1}{2} \tag{5.86}$$

Evaluate the accuracy of the approximation.

8. Solve the following SDP using Python

$$\min x + y \tag{5.87}$$

$$\begin{pmatrix} x & 1 \\ 1 & y \end{pmatrix} \geq 0 \tag{5.88}$$

$$x + y \leq 3 \tag{5.89}$$

Plot the feasible set, formulate and solve the dual problem, and analyze the results.

9. Semidefinite programming models can be solved by interior point methods. In this case, a constraint of the form $X \geq 0$ can be penalized by a barrier function given by $\phi(X) = -\mu \ln(\det(X))$. Use this fact to solve the following feasibility problem:

$$\min 0$$

$$X \geq 0 \tag{5.90}$$

$$\text{tr}(X) = 1$$

The following equations can be useful:

$$\frac{\partial \ln(\det(X))}{\partial X} = X^{-1}$$

$$\frac{\partial \operatorname{tr}(AX)}{\partial X} = A \tag{5.91}$$

where both A and X are square matrices.

10. Generate a random instance of the problem presented in Example 5.12 for $Q \in \mathbb{R}^{1}0$. Find the optimal solution using a brute-force search and a semidefinite approximation. Compare the solution.

6

Robust optimization

Learning outcomes

By the end of this chapter, the student will be able to:

- Represent uncertainties through an uncertainty set.
- Formulate simple, robust optimization problems.

6.1 Stochastic vs robust optimization

So far we have been concerned with the problems of the following form:

$$\min\ f(x,\beta) \tag{6.1}$$

Where $x \in \mathbb{R}^n$ is a vector of decision variables, and β is a vector of parameters. However, it is common that β is not completely known. Hence we require to decide in the presence of uncertainty. There are two classical ways to deal with this problem, namely: stochastic optimization and robust optimization. In stochastic optimization, it is assumed that the probability distribution of β is known. We need extensive statistical data to define this distribution. In practice, this information may not be available. In robust optimization, we require less information since we only assume that β is in a closed uncertainty set $\mathcal{U}$. In this case, our goal is to find an optimal solution in the sense of f and, at the same time, a feasible solution regardless of the value of $\beta \in \mathcal{U}$. We want the best solution in the worst-case scenario. Below, we present a naive introduction to stochastic optimization, and next, we present the robust approach.

Mathematical Programming for Power Systems Operation: From Theory to Applications in Python. First Edition. Alejandro Garcés.
© 2022 by The Institute of Electrical and Electronics Engineers, Inc. Published 2022 by John Wiley & Sons, Inc.

6.1.1 Stochastic approach

There are several ways to deal with randomness and risk in stochastic optimization. One of the strategies, called *here and now*, assumes that the optimization problem is solved at the beginning of the planning horizon, taking into account future uncertainty. The second strategy is called *wait and see*. In this strategy, we make some decisions at the first stage and make modifications throughout the planning horizon. In both cases, we require a suitable model of the stochastic process.

A common objective for stochastic problems is to minimize an expect value, as presented below:

$$\min \; \mathbb{E}(f(x, \beta))$$
$$\beta \in \mathcal{M} \tag{6.2}$$

where $\mathcal{M}$ represents the probability space in which β lives. There are many ways to represent $\mathcal{M}$. For example, it may be represented as a set of discrete scenarios with a given probability, or as a continuous distribution function. In any case, the challenge is not only the representation of the uncertainty but the tractability of the resulting optimization model.

6.1.2 Robust approach

In a robust optimization problem, we have a continuous uncertainty set $\mathcal{U}$ that contains β. This set plays a similar role than $\mathcal{M}$ in the stochastic approach. However, we do not have additional information related to the probability, i.e., we know that β belongs to $\mathcal{U}$, but we ignore which regions are more probable. Therefore, we can define an optimization problem that seeks to minimize f under the worst-case scenario of β, as given in (6.3),

$$\min_{x} \left\{ \sup_{\beta \in \mathcal{U}} f(x, \beta) \right\} \tag{6.3}$$

A critical step in the problem above is defining the uncertainty set $\mathcal{U}$ to obtain a tractable and realistic problem. In the following sections, we present how to define this set and how it is related to the objective function and the problem's constraints. We use the following simple linear programming problem to present the main concepts:

$$\min \; c^{\mathsf{T}} x$$
$$a^{\mathsf{T}} x \leq b \tag{6.4}$$

where c, a, and/or b may be uncertain parameters of the model.

6.2 Polyhedral uncertainty

Let us consider the case in which a is contained in a polyhedral uncertainty set $\mathcal{U} \subset \mathbb{R}^n$ defined as follows:

$$\mathcal{U} = \{a : D^\mathsf{T} a \leq d\} \tag{6.5}$$

The problem seems highly complex since both a and x are unknown. However, the problem can be solved in two step: first, we determine a model with a worst-case outcome and then, we optimize this model. The worst-case outcome for (6.4) is given below:

$$\left(\sup_{a \in \mathcal{U}} a^\mathsf{T} x \right) \leq b \tag{6.6}$$

Then, the following optimization problem is raised:

$$\mathcal{P}(a) = \left\{ \sup_{a \in \mathcal{U}} a^\mathsf{T} x \right\} \tag{6.7}$$

This is a linear programming problem, where the decision variables are a, and the uncertainty set $\mathcal{U}$ is a polytope given by (6.5). Therefore, we can define a primal problem $\mathcal{P}(a)$ written as follows:

$$\mathcal{P}(a) = \left\{ \begin{array}{c} \max\ a^\mathsf{T} x \\ D^\mathsf{T} a \leq d \end{array} \right\} \tag{6.8}$$

Next, we define the dual problem $\mathcal{D}(y)$, as given below[1]:

$$\mathcal{D}(y) = \left\{ \begin{array}{c} \min\ y^\mathsf{T} d \\ y^\mathsf{T} D = x \\ y \geq 0 \end{array} \right\} \tag{6.9}$$

The following linear programming problem is obtained, after replacing into (6.4):

$$\begin{aligned} \min\ & c^\mathsf{T} x \\ & y^\mathsf{T} d \leq b \\ & y^\mathsf{T} D = x \\ & y \geq 0 \end{aligned} \tag{6.10}$$

The previous analysis considered only data uncertainty in the constraints. However, the objective function may also be subject to uncertainty. In that case,

---

1 The reader is invited to review Section 3.5 for studying the basic duality theory.

it is convenient to transform (6.4) into the following equivalent representation:

$$\min z$$
$$c^T x \leq z \tag{6.11}$$
$$a^T x \leq b$$

Thus, the uncertainty set must now include the values of c. The following example helps to understand this model.

Example 6.1. Let us consider the following optimization problem:

$$\min 3x_0 + 5x_1$$
$$x_0 + x_1 \geq 1 \tag{6.12}$$
$$x_0, x_1 \geq 0$$

The solution to this problem is $x = (1,0)$ with objective function $z = 3$. Let us suppose now, the coefficients in the objective function are uncertain, with a maximum deviation of ± 0.5. The following polyhedral uncertainty is defined:

$$\mathcal{U} = \{(c_0, c_1) : 2.5 \leq c_0 \leq 3.5, 4.5 \leq c_1 \leq 5.5\} \tag{6.13}$$

We define the worst-case outcome as follows:

$$\min z$$
$$\left(\sup_{(c_0, c_1) \in \mathcal{U}} c_0 x_0 + c_1 x_1 \right) \leq z$$
$$x_0 + x_1 \geq 1 \tag{6.14}$$
$$x_0, x_1 \geq 0$$

The primal problem $\mathcal{P}$ associated to the uncertainty set is given below:

$$\max c_0 x_0 + c_1 x_1$$
$$\begin{pmatrix} 1 & 0 \\ -1 & 0 \\ 0 & 1 \\ 0 & -1 \end{pmatrix} \begin{pmatrix} c_0 \\ c_1 \end{pmatrix} \leq \begin{pmatrix} 3.5 \\ -2.5 \\ 5.5 \\ -4.5 \end{pmatrix} \tag{6.15}$$

Now, we formulate the dual model to obtain a robust equivalent, namely:

$$\min z$$
$$3.5y_0 - 2.5y_1 + 5.5y_2 - 4.5y_3 \leq z$$
$$y_0 - y_1 = x_0$$
$$y_2 - y_3 = x_1 \tag{6.16}$$
$$x_0 + x_1 \geq 1$$
$$x_0, x_1 \geq 0$$
$$y_0, y_1, y_2, y_3 \geq 0$$

The solution to this problem is $z = 3.5$. This solution is, indeed, worst than the solution of (6.12). However, it is the best solution in the worst-case scenario, i.e., it is a robust solution.

6.3 Linear problems with norm uncertainty

Uncertainty may also be represented by a closed ball $\mathcal{B}$ in $\mathbb{R}^n$. In that case, all coefficients a are uncertain. However, we might know they are inside a ball with center in α and radius δ, as follows:

$$\min c^\top x$$
$$a^\top x \leq b \tag{6.17}$$
$$a \in \mathcal{B} = \{a \in \mathbb{R}^n : a = \alpha + \delta \xi, \text{with}, \|\xi\| \leq 1\}$$

where $\|\cdot\|$ is any norm in $\mathbb{R}^n$. Likewise the polyhedral uncertainty, we require to determine the wort-case outcome, is given below:

$$\left(\sup_{a \in \mathcal{B}} a^\top x\right) \leq b \tag{6.18}$$

However, this is not a linear programming problem, since the set of feasible solutions, $\mathcal{B}$, is non-linear. Therefore, the following optimization problem is raised:

$$\max \alpha^\top x + \delta \xi^\top x$$
$$\|\xi\| \leq 1 \tag{6.19}$$

Table 6.1 Dual norms for the most common cases.

Norm	Dual norm
norm-1	norm-∞
norm-2	norm-2
norm-∞	norm-1

where the decision variables are ξ. We can formulate the dual model associated with this problem and proceed in the same way as in the previous cases. Nevertheless, a more systematic way to solve the problem is by defining a new function called dual norm, as follows:

$$\|y\|_{\mathcal{D}} = \sup\{y^\top x, \ \|x\| \le 1\} \tag{6.20}$$

This norm holds all properties presented in Section 2.2. In addition, it is a bijective operation, which means that the dual norm of $\|\cdot\|_{\mathcal{D}}$ is again $\|\cdot\|$. Table 6.1 shows the dual norms of the three most common cases.

With this useful definition, we can easily formulate a tractable model for (6.17), namely:

$$\min c^\top x$$
$$\alpha^\top x + \delta \|x\|_{\mathcal{D}} \le b \tag{6.21}$$

notice this is a convex optimization problem since a norm is a convex function. The problem might be reduced to a linear programming problem for the cases of 1-norm and $\infty - $norm. It is a second-order cone optimization problem for the case of the 2-norm.

Example 6.2. Let us consider the following linear programming problem:

$$\min z = -8x_0 - 7x_1 - 9x_2$$
$$x_0 + x_1 + x_2 \le 10 \tag{6.22}$$
$$x \ge 0$$

This problem has an optimum in $x = (0, 0, 10)^\top$ with $z = -90$. Now, let us consider the case in which the coefficients associated to the first constraint are $a = (a_0, a_1, a_2)^\top$ with

$$(a_0 - 1)^2 + (a_1 - 1)^2 + (a_2 - 1)^2 \le 0.1 \tag{6.23}$$

This constraint is equivalent to say that $a = (1, 1, 1)^\top + \xi$ with $\|\xi\| \le 0.3162$. Therefore, the equivalent robust optimization problem is given by the following model:

$$\min \ -8x_0 - 7x_1 - 9x_2$$

$$x_0 + x_1 + x_2 + 0.3162\,\|x\|_2 \le 10 \tag{6.24}$$

$$x \ge 0$$

The optimal solution of this problem is $z = -70.1356$ with $\tilde{x} = (2.685, 0, 5.406)^\top$.

6.4 Defining the uncertainty set

The uncertainty set can be defined using statistic information of the parameters. For example, it might be the case that $a \in \mathbb{R}$ is a single variable which is normally distributed, i.e., $a \sim \mathcal{N}(\alpha, \sigma)$, where α is the mean and σ is the standard deviation. The probability density function of a is presented below:

$$\psi(a) = \frac{1}{\sigma\sqrt{2\pi}} \exp\left(-\frac{1}{2}\left(\frac{a-\alpha}{\sigma}\right)^2\right) \tag{6.25}$$

We can define a confidence interval $\mathcal{U}$ for a. Hence, the probability that a belongs to $\mathcal{U}$ is given by (6.26),

$$\text{Prob}\,(a \in \mathcal{U}) = \int_{\mathcal{U}} \psi(a)da \tag{6.26}$$

With this simple approach, we can define the uncertainty set for a single parameter a. The main idea is to replace the constraint in (6.4) by a chance constraint as follows:

$$\text{Prob}(ax \le b) \ge \eta \tag{6.27}$$

where η is the probability given by (6.26). This equation implies that the probability of meeting the constraint is above a given value η.

Example 6.3. Figure 6.1 shows the probability density function for a parameter a with mean $\alpha = 20$ and standard deviation $\sigma = 1$. The set $\mathcal{U}$ constitutes a confidence interval for the robust optimization problem. Two uncertainty sets are defined in Figure 6.1, namely: $\mathcal{U}_A = [19, 21]$ and $\mathcal{U}_B = [18, 22]$. The set $\mathcal{U}_A$ includes values between $\pm\sigma$ with a probability $\text{Prob}(a \in \mathcal{U}_A) = 68\%$, while the set $\mathcal{U}_B$ includes values between $\pm 2\sigma$ and its probability is $\text{Prob}(a \in \mathcal{U}_B) = 95\%$. These probabilities were calculated using (6.26).

This same idea may be applied for a in higher dimensions, for example, $a \in \mathbb{R}^n$. In that case, the univariate normal distribution is replaced for a multivariate normal distribution with the following probability density function:

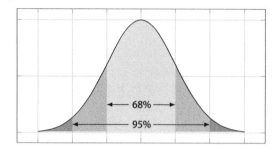

Figure 6.1 Probability density function for a variable with mean $\alpha = 20$ and standard deviation $\sigma = 1$

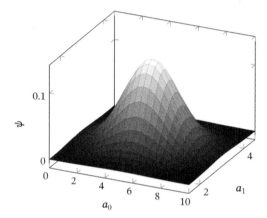

Figure 6.2 A multivariate normal distribution in $\mathbb{R}^2$

$$\psi(a) = \frac{1}{\sqrt{(2\pi)^n \det(S)}} \exp\left(-\frac{1}{2}(a - \alpha)^\mathsf{T} S^{-1}(a - \alpha)\right) \tag{6.28}$$

where α is a vector of mean values, and S is a positive definite covariance matrix. This multivariate distribution can be used to define the uncerntanty set as presented in the next example.

Example 6.4. Figure 6.2 depicts a probability density function for $\mathbb{R}^2$. The confidence intervals are now replaced by the following confidence regions,

$$\mathcal{U}_\gamma = \left\{a \in \mathbb{R}^2 : (a - \alpha)^\mathsf{T} S^{-1}(a - \alpha) \leq \gamma\right\} \tag{6.29}$$

These confidence regions are ellipsoid as depicted in Figure 6.3

Example 6.5. The probability associated to a confidence region $\mathcal{U}_\gamma$ can be calculated using a numerical approach, based on Monte Carlo simulation.Let

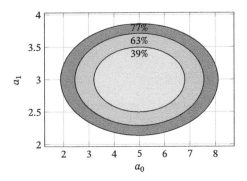

Figure 6.3 Confidence regions for a multivariate normal distribution in $\mathbb{R}^2$

us consider a multivariate normal distribution with $\alpha = (5,3)$ and $S = \text{diag}(3.24, 0.25)$. We use the function Multivariate Normal from the module SciPy to generate n_{points} random scenarios of a; next, we count the number of scenarios in which $a \in \mathcal{U}_\gamma$; this number is stored in a variable η_{in}; the probability $\text{Prob}(a \in \mathcal{U}_\gamma)$ is given by $\eta_{\text{in}}/n_{\text{points}}$. This approach is more precise for a high value of n_{points}. The code in Python is presented below:

```
import numpy as np
from scipy.stats import multivariate_normal
alpha = np.array([5,3])
S = np.array([[3.24,0],[0,0.25]])
p = multivariate_normal(alpha,S)
n_points = 10000
Sinv = np.linalg.inv(S)
gamma = 5
eta_in = 0
for k in range(n_points):
    a = p.rvs()
    w = np.array(a-alpha)
    z = w.T@Sinv@w
    if z <= gamma: eta_in += 1
print('Prob = ',eta_in/n_points)
```

Notice this method can be applied to any distribution in $\mathbb{R}^n$.

A linear constraint $a^\mathsf{T} x \leq b$ with $a \sim \mathcal{N}(\alpha, S)$ can be transformed into a robust optimization problem by defining a confidence ellipsoid given by (6.29). Without lost of generality, we define a new parameter $\delta \sim \mathcal{N}(0, S)$; therefore, the linear constraint takes the following form:

$$\alpha^\mathsf{T} x + \sup_{\delta \in \mathcal{U}_\gamma} \left(\delta^\mathsf{T} x\right) \le b \tag{6.30}$$

with

$$\mathcal{U}_\gamma = \{\delta \in \mathbb{R}^n : \delta^\mathsf{T} S^{-1} \delta \le \gamma^{1/2}\} \tag{6.31}$$

This set can also be defined using a second-order cone, as presented below:

$$\mathcal{U}_\gamma = \{\delta \in \mathbb{R}^n : \gamma^{-1/2} \left\| S^{-1/2}\delta \right\| \le 1\} \tag{6.32}$$

where $S^{-1/2}$ is the Cholesky decomposition of S^{-1}. This decomposition exists since S is positive definite. Let us define a new variable $z = \gamma^{-1/2} S^{-1/2}\delta$, then, the robust constraint can be represented as follows:

$$\alpha^\mathsf{T} x + \sup_{\|z\| \le 1} \left(\gamma^{1/2} z^\mathsf{T} S^{1/2} x\right) \le b \tag{6.33}$$

The supreme in the left-hand side of (6.33) can be represented as the dual norm of norm-2. Consequently, the robust constraint is defined by the following second-order cone:

$$\alpha^\mathsf{T} x + \gamma^{1/2} \left\| S^{1/2} x \right\| \le b \tag{6.34}$$

This is, evidently, a convex constraint.

Example 6.6. Let us consider the following linear programming problem:

$$\min q = -10x_0 - 15x_1$$
$$a_0 x_0 + a_1 x_1 \le 10 \tag{6.35}$$
$$x_0, x_1 \ge 0$$

Where $a = (a_0, a_1)^\mathsf{T}$ is normally distributed with mean $\alpha = (5, 3)$ and covariance matrix $S = \text{diag}(3.24, 0.25)$ (the same parameters of Example 6.5). The solution to this linear programming problem for $a = \alpha$ is $q = -50$ and $x = (0, 10/3)^\mathsf{T}$. The robust solution can be calculated with different degree of robustness. For instance, for $\gamma = 5$ we have a 92% probability to hold the constraint. The equivalent robust optimization problem is presented below:

```
import numpy as np
import cvxpy as cvx
c = np.array([-10,-15])
alpha = np.array([5,3])
S = np.array([[3.24,0],[0,0.25]])
M = np.linalg.cholesky(S)
r = np.sqrt(1/5)
x = cvx.Variable(2, nonneg=True)
```

```
obj = cvx.Minimize(c.T@x)
res = [cvx.SOC(r*10-r*alpha.T@x,M@x)]
Model = cvx.Problem(obj,res)
Model.solve(verbose=True)
print(np.round(x.value,3))
print(np.round(obj.value))
```

The solution of this problem is $q = -36$ and $x = (0, 2.428)^\mathsf{T}$. This solution is clearly lower than the solution of the base problem. However, this solution is robust enough to guarantee that the solution is feasible in 92% of the scenarios.

Another simple but common case of robust optimization, is when the uncertainty is associated to b in (6.4). In that case, the constraint may be transformed into a robust problem as presented below:

$$a^\mathsf{T} x \leq \phi^{-1}(\eta) \tag{6.36}$$

where ϕ_b^{-1} is the quantile function[2] associated to the distribution of b, and η is the probability to hold the constraint. It is not required that this distribution is normal.

Example 6.7. Let us consider the following constraint:

$$8x + 15y \leq b \tag{6.37}$$

where b is normally distributed with mean $\mu = 10$ and standard deviation $\sigma = 1$. We can see the histogram for this parameter by generating a high number of random scenarios and using the corresponding function in MatplotLib, as follows:

```
import numpy as np
import matplotlib.pyplot as plt
b = 10 + np.random.randn(10000)
plt.hist(b,20)
plt.grid()
plt.show()
```

The quantile with a given probability is obtained using the quantile function, as presented below:

---

2 The quantile function is defined as $\phi_b^{-1}(p) = \inf\{b \in \mathbb{R} : F(b) \geq p\}$; F is the cumulative distribution function.

```
print('Quantile at 98%: ',np.quantile(b,1-0.98))
```

This function results in a value of $b = 7.9$. Therefore, the robust constraint associated to (6.37) is (6.38),

$$8x + 15y \leq 7.97 \tag{6.38}$$

Example 6.8. The numerical method presented in the previous example can be extended to obtain robust solution in problems where b has a distribution different from the normal distribution. That is the case of the wind velocity which is often approximated by the Weibull distribution. However, the distribution of the generated power is different, since the output power of a wind turbine depends on its control. Usually, a wind turbine is controlled to obtain maximum efficiency for wind velocities between 0 and v_{nom}; consequently the turbine is controlled to obtain nominal power; finally, for wind velocities higher than v_{max}, the turbine is blocked. This control can be represented by the following equation:

$$p(v) = \begin{cases} p_{nom} \left(v/v_{nom} \right)^3 & 0 \leq v \leq v_{nom} \\ p_{nom} & v_{nom} < v \leq v_{max} \\ 0 & v > v_{max} \end{cases} \tag{6.39}$$

Let us consider a wind turbine with $v_{nom} = 12$ and $v_{max} = 25$. This turbine is located in an offshore emplacement where the wind varies according to a Weibull distribution with scale factor $\lambda = 13$ and shape $a = 2$. A histogram of this variable can be obtained as follows:

```
import numpy as np
import matplotlib.pyplot as plt
v = 13*np.random.weibull(2, 10000)
plt.hist(v,20)
plt.grid()
plt.show()
```

However, we may be interested in the distribution of the output power; therefore, we define (6.39) as a Python function, as presented below:

```
def wind_power(w):
    p = 0
    if (w>0)&(w<=12):
        p = 2*(w/12)**3
    if (w>12)&(w<=25):
        p = 2
    return p
pt = np.zeros(len(v))
```

```
for k in range(len(v)):
  pt[k] = wind_power(v[k])
plt.hist(pt)
plt.grid()
```

Finally, we can evaluate the quantile function for different probabilities, namely:

```
q = []
for k in range(100):
    q += [np.quantile(pt,1-k/100)]
plt.plot(q)
plt.grid()
```

This plot allows obtaining different quantiles according to the expected probability.

6.5 Further readings

Robust optimization is a rich area of research with many impressive theoretical results. Several applications can be found in scientific journals; for example, an application for smart grids can be found in [33]. A general review of robust optimization is available in [34].

Another approach to deal with the uncertainty in optimization problems is using stochastic optimization. A complete analysis of this approach is beyond the objectives of this book. An excellent presentation of this subject is given in [35].

6.6 Exercises

1. Demonstrate the relation between the norm and dual norm for the cases shown in Table 6.1. Use the definition of dual norm, given by (6.20), and the theory presented in Section 2.2.
2. Consider a norm $\|x\|$ and its corresponding dual norm $\|x\|_{\mathcal{D}}$. Show that these norms holds the following inequality:

$$x^T y \le \|x\| \, \|y\|_{\mathcal{D}} \tag{6.40}$$

3. Example 6.2 used norm-2 to define the uncertainty set. Formulate and solve the same problem but now using norm-1 and norm-∞.

4. Consider a matrix $H = H^\top > 0$ and the function

$$n(x) = x^\top H x \tag{6.41}$$

show that $n(x)$ is a norm and determine its dual norm.

5. Consider the following optimization problem

$$\min f(x)$$
$$a^\top x \le b \tag{6.42}$$

where a is a vector contained in the ellipsoid $\mathcal{E} = \{a^\top H a \le 1\}$, with $H > 0$.

6. Solve the problem in Example 6.6 for different values of γ. Show plots of q vs γ and Probability vs γ.

7. Repeat the calculations of Example 6.5 for $a \in \mathbb{R}^3$; consider a normal distribution $a \sim \mathcal{N}(\alpha, S)$ with $\alpha = (15, 12, 25)^\top$ and S given by the following matrix:

$$S = \begin{pmatrix} 1.0 & 0.5 & 0.1 \\ 0.5 & 2.0 & 0.1 \\ 0.1 & 0.1 & 2.0 \end{pmatrix} \tag{6.43}$$

8. Find a robust counterpart for the following optimization problem:

$$\min ax + by + cz$$
$$x + y + z = 100$$
$$0 \le x \le \delta \tag{6.44}$$
$$0 \le y \le \delta$$
$$0 \le z \le \delta$$

with $\delta = 50$; $a \sim \mathcal{N}(30, 4)$, $b \sim \mathcal{N}(31, 2)$, and $c \sim \mathcal{N}(32, 1)$;

9. Solve the previous problem but now $\delta \sim \mathcal{N}(50, 3)$.

10. Robust optimization tends to be too conservative since robust solutions may occur when all the parameters deviate simultaneously to the worst condition. This condition, although robust, may be unlikely in practice. One way to qualify the solution is by using cardinality constrained uncertainty. In this approach, the uncertainty set is represented as a polyhedron; however, we shall allow at must Γ coefficients to deviate. Let us consider Model (6.4) with $a = \bar{a} \pm \delta$, with the following primal problem:

$$\max \delta^\mathsf{T} v$$

$$v_i = |x_i| w_i$$

$$\sum w_i \le \Gamma \tag{6.45}$$

$$0 \le w_i \le 1$$

Formulate the dual problem associated with (6.45) and the robust counterpart of the original problem.

Part II

Power systems operation

7

Economic dispatch of thermal units

Learning outcomes

By the end of this chapter, the student will be able to:

- Formulate the problems of economic and environmental dispatch.
- Include constraints related to the active power loss.
- Include constraints related to the transmission lines' capacity considering the transportation model or the linear power flow.

7.1 Economic dispatch

The economic dispatch of thermal units was one of the first mathematical programming applications to power systems operation. Historically, the first implementations of economic dispatch models coincided with the computer development, which allowed to make automatic calculations efficiently and in real-time [36]. The problem consists of determining the most economical manner of operation to supply a given load condition. Each thermal power plant has a different relationship between input (i.e., fuel) and output (i.e., electric power) according to the type of fuel, thermodynamic cycle, and particular plant characteristics. Therefore, a cost function is defined for each thermal unit. Figure 7.1 depicts schematically, the economic dispatch problem for three thermal units supplying a single load. In its most basic form, the effect of the grid is neglected leading to an optimization problem as given in Equation (7.1),

Mathematical Programming for Power Systems Operation: From Theory to Applications in Python. First Edition. Alejandro Garcés.
© 2022 by The Institute of Electrical and Electronics Engineers, Inc. Published 2022 by John Wiley & Sons, Inc.

$$\min \sum_{k \in \mathcal{J}} f_k(p_k)$$

$$\sum_{k \in \mathcal{J}} p_k = d \tag{7.1}$$

where $\mathcal{J}$ is the set of thermal units, f_k is the cost function for each unit $k \in \mathcal{J}$, p_k is the generated power, and d is the total demand. In this model, a number of simplifications were made on the manner in which power systems would be operated. For instance, power losses, grid constrains, and capacity of the generation units were neglected (these aspects will be considered later on in this chapter). Model Equation (7.1) can be solved by the method of Lagrange multipliers with the following Lagrangian:

$$\mathcal{L}(p, \lambda) = \sum_{k \in \mathcal{J}} f_k(p_k) + \lambda(d - \sum_{k} p_k) \tag{7.2}$$

The first-order condition for optimal solution is obtained by deriving $\mathcal{L}$ with respect to p_k:

$$\frac{\partial f_k}{\partial p_k} - \lambda = 0 \tag{7.3}$$

The value of $\partial f_k / \partial p_k$ is known as *incremental cost*. Therefore, the optimal dispatch is obtained when the incremental costs of all thermal units are the same. The second condition is obtained by deriving $\mathcal{L}$ with respect to λ and gives the power balance (i.e., the sum of the generation must be equal to the demand).

Cost functions are usually represented as quadratic function as given in Equation (7.4),

$$f_k(p_k) = \frac{a_k}{2} p_k^2 + b_k p_k + c_k \tag{7.4}$$

where a_k, b_k, c_k are constants fit from data of the input to output relation of each thermal unit. A quadratic representation of the thermal units simplifies the problem enormously. The optimal conditions are the following set of linear equations that can be easily solved in practice:

$$a_k p_k + b_k = \lambda \tag{7.5}$$

$$\sum_{k \in \mathcal{J}} p_k = d \tag{7.6}$$

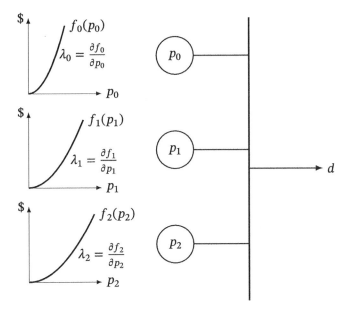

Figure 7.1 Three thermal units with their respective cost functions for the economic dispatch problem.

Actual power units have limits of minimum and maximum generation ($p^{\min}$, $p^{\max}$) that must be included into the model as follows:

$$\min \sum_{k \in \mathcal{J}} f_k(p_k)$$

$$\sum_{k \in \mathcal{J}} p_k = d \qquad (7.7)$$

$$p_k^{\min} \leq p_k \leq p_k^{\max}, \ \forall k \in \mathcal{J}$$

The effect of these inequality constraints is shown in Figure 7.2. We may obtain the most economical dispatch by solving a set of linear equations if the solution is within the operating limits (that is the case for λ_A). However, one unit may achieve full load before the others, as is the case of Unit 2 for the incremental cost λ_B. In that case, Unit 2 is set to the maximum ($p_2 = p^{\max}$) and the rest of the demand is supplied by the other two units, at equal incremental cost. In general, the problem is solved as a quadratic optimization problem as presented in the following examples.

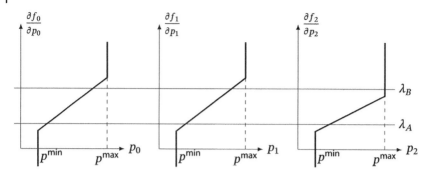

Figure 7.2 Incremental cost for three thermal units considering capability limits.

Example 7.1. Let us consider a system with two thermal units that supply a demand of $d = 200$ MW, where the costs functions of each unit is given below:

$$f_0(p_0) = \frac{0.31}{2} p_0^2 + 38p_0 \tag{7.8}$$

$$f_1(p_1) = \frac{0.22}{2} p_1^2 + 46p_1 \tag{7.9}$$

Our objective is to supply demand at the minimum cost. Therefore, we require to minimize the following Lagrangian equation:

$$f_0(p_0) + f_1(p_1) + \lambda(d - p_0 - p_1) \tag{7.10}$$

which results in the following set of linear equations:

$$0.31p_0 + 38 - \lambda = 0 \tag{7.11}$$

$$0.22p_1 + 46 - \lambda = 0 \tag{7.12}$$

$$p_0 + p_1 = 200 \tag{7.13}$$

The solution of this linear system was $p_0 = 98.11$, $p_1 = 101.88$ and $\lambda = 68.42$. The code in Python for solving this simple problem, is presented below:

```
A = [[0.31,0,-1],[0,0.22,-1],[1,1,0]]
b = [-38,-46,200]
x = np.linalg.solve(A,b)
print(x)
```

This problem was simple enough to be solved without any optimization solver. However, as the number of variables increases, the problem becomes more complicated. In addition, the presence of box constraints for the maximum and minimum power generation, makes necessary the use of a general quadratic programming solver.

Table 7.1 Cost functions and operative limits for a system with six thermal units [37].

Unit	P_{min} (MW)	P_{max} (MW)	a_k ($/MWh^2)	b_k ($/MWh)	α_k (lb/MWh^2)	β_k (lb/MWh)
T0	10	125	0.30494	38.5390	0.00838	0.32767
T1	10	150	0.21174	46.1591	0.00838	0.32767
T2	35	210	0.07092	38.3055	0.01366	−0.54551
T3	35	225	0.05606	40.3965	0.01366	−0.54551
T4	125	315	0.03598	38.2704	0.00922	−0.51116
T5	130	325	0.04222	36.3278	0.00922	−0.51116

Example 7.2. Consider a system with six units with parameters given in Table 7.1 and demand $d = 1200$ MW.

For the sake of simplicity, the values of c_k were set to zero[1]. The economic dispatch consists on a quadratic programming problem, given by Equation (7.7), that can be coded in Python as follows:

```python
import numpy as np
import cvxpy as cvx
pmin = [10,10,35,35,125,130]
pmax = [125,150,210,225,315,325]
a = np.diag([0.30494,0.21174,0.07092,0.05606,0.03598,0.04222])
b = [38.5390,46.1591,38.3055,40.3965,38.2704,36.3278]
d = 1200
p = cvx.Variable(6)
obj = cvx.Minimize(1/2*cvx.quad_form(p,a)+b*p)
res = [sum(p) >= d , p>=pmin, p<=pmax]
Model = cvx.Problem(obj,res)
Model.solve()
print(p.value)
print('Incremental Cost:',res[0].dual_value)
```

The economic dispatch for this case is $p = (66, 59, 210, 225, 315, 325)^\top$ and the incremental cost is $\lambda = 58.66$ (the reader is invited to implement the code to prove the results). Notice that the objective function is convex since $a_k \geq 0$. In addition, the function is strictly convex, therefore, this is the global and unique optimum of the problem.

Example 7.3. Linear models are common for economic dispatch problem in modern electricity markets. In that case, the optimization model is highly simplified, as presented below:

---

1 Parameters α, β will be used in other examples below.

$$\min \sum_k c_k p_k$$

$$\sum_k c_k = d \tag{7.14}$$

$$0 \le p \le p_k^{\max}$$

The following heuristic algorithm, known as merit order method, can solve this linear problem: First, all units are organized according to the price c_k, from the unit with minimum cost to the unit with maximum cost (i.e., ascending order of price). Next, each unit is dispatched with its maximum power until the total demand is supplied. The spot price is the dual variable associated to the power balance constraint.

Example 7.4. A linear model might be obtained from the quadratic cost as given in Figure 7.3. A linear approximation enormously simplifies the optimization model.

In that case, each thermal unit is represented by a linear cost function $\zeta_k p_k$, where the value of ζ_k can be calculated as follows:

$$\min \int_0^{p^{\max}} \left(\zeta_k p_k - \frac{a_k}{2} p_k^2 - b_k p_k \right)^2 dp_k \tag{7.15}$$

This is a simple minimum square problem with the following solution:

$$\zeta_k = \frac{3a_k}{8} p^{\max} + b_k \tag{7.16}$$

The economic dispatch is transformed into a linear programming model with this approximation. Example 7.2 was solved with this model obtaining the following result: $p = (115, 10, 210, 225, 315, 325)^\top$. Notice the linear model agreed

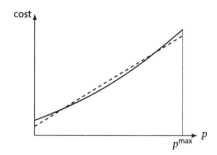

Figure 7.3 Linear approximation of quadratic cost functions.

with the quadratic model in p_3 to p_5 but it was not accurate for p_0 and p_1. However, total costs were not that different in one and the other case. 🐍

Example 7.5. It might be the case that c_k in Equation (7.14), is unknown, but determined by a temporal series, i.e, $c_k = \bar{c}_k \pm \delta_k$, where δ_k is normally distributed with mean $\mu = 0$ and standard deviation δ_k. In that case, we can obtain a robust optimization model for the economic dispatch problem. The robust optimization problem is the one presented below:

min z

$$\sum_k \bar{c}_k p_k + \gamma^{1/2} \left\| \sum_k \sigma_k p_k \right\| \leq z$$

$$\sum_k c_k = d \tag{7.17}$$

$$0 \leq p \leq p_k^{max}$$

where γ defines the size of the confidence region[2]. The second-order cone in the first constraint can be interpreted as a penalization factor for a deviation of the cost c_k. A high penalization factor results in a robust although perhaps not very efficient solution. A trade-off between cost and robustness can be defined by this parameter. 🐍

Example 7.6. Uncertainty in the load or renewable power generation (e.g., solar and wind) can be introduced in the model by using robust optimization. In that case, we just change the power balance for the following robust constraint:

$$\sum_k p_k \leq \phi_d^{-1}(\eta) \tag{7.18}$$

where ϕ_d^{-1} is the quantile function of the load d, and η is the desired probability.

🐍

7.2 Environmental dispatch

Another problem closely related to economic dispatch is environmental dispatch. In this problem, the cost function is replaced by an emission function that considers greenhouse gas emissions such as sulfur oxides (SOx) and nitrogen oxides (NOx). The former has a quadratic form, whereas the latter is

---

2 See Chapter 6.

usually characterized by an equation consisting of a straight line and an exponential function. Thus, the environmental dispatch has the same structure as Equation (7.7) with emission functions given by Equation (7.19).

$$f_k(p_k) = \frac{\alpha_k}{2} p_k^2 + \beta_k p_k + \gamma_k \exp(\eta_k p_k) \tag{7.19}$$

Notice this function is convex if $\alpha_k \geq 0$. The analysis of this type of problem is straightforward.

Example 7.7. Solve the environmental dispatch problem for a power system with six units presented in Table 7.1 where only SOx emissions are considered (i.e., the objective function is quadratic). The script for solving the problem is the same as Example 7.2 replacing the values of a_k for α_k and b_k for β_k. The solution is $p = (125, 150, 188, 188, 275, 275)^T$, notice this solution is different from the economic dispatch since economic and environmental objectives are usually contradictory. For example, the unit $T0$ was dispatched with less power in the economic dispatch than the environmental dispatch since it is more expensive but less pollutant.

The economic dispatch and the environmental dispatch are usually contradictory objectives. Therefore, it is required to study the problem as a multiobjective optimization. Although there is a trend in the scientific literature of power systems for using heuristic algorithms in multiobjective optimization problems, the economic/environmental dispatch has convex objectives that allow an immediate solution, as presented below.

Consider two optimization problems with the same set of feasible solutions but two contradictory objectives, f_A and f_B; this situation can be represented as Equation (7.20).

$$\min \{f_A(x), f_B(x)\}$$
$$f(x) = 0 \tag{7.20}$$
$$g(x) \leq 0$$

Solving the problem concerning f_A may lead to an unacceptable solution regarding f_B and vice versa. Therefore, it is required to find a trade-off between the two objectives. We can find this trade-off through the concept of the Pareto frontier. Consider three feasible solutions A, B, C depicted in Figure 7.4 for a two-objective optimization problem. Solution A is better than B regarding f_A, but B is better than A regarding f_B. However, no solution is better in both objectives simultaneously. This type of solution is named as non-dominated solution. Instead, C is a dominant solution since there are better solutions in both objectives (i.e., solutions A and B are better than C in both objectives). The set of non-dominated solutions is called the Pareto frontier, and the multiobjective

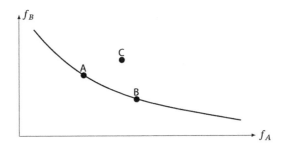

Figure 7.4   Example of a Pareto frontier for two contradictory objective functions that require to be minimized.

problem is considered solved when the Pareto frontier is found. The final decision about the dispatch is made by the transmission system operator using this frontier; for example, the transmission system operator may decide the maximum deviation of the economic dispatch that it is willing to accept to reduce emissions.

The Pareto frontier can be found by transforming the multiobjective problem into a single-objective problem as follows:

$$\min \xi f_A(x) + (1 - \xi)f_B(x)$$
$$f(x) = 0 \qquad\qquad\qquad (7.21)$$
$$g(x) \le 0$$

where ξ is a real number between 0 and 1. This model is convex since both f_A and f_B are convex. The Pareto frontier is obtained by solving the problem for different values of ξ as presented in the example below:

Example 7.8. Consider the economic/enviromental dispatch for the system presented in Table 7.1. In this case, it is required to define the single-objective model Equation (7.21) and solve for different values of ξ between 0 and 1 as follows:

```
import numpy as np
import cvxpy as cvx
import matplotlib.pyplot as plt
pmin = [10,10,35,35,125,130]
pmax = [125,150,210,225,315,325]
a = np.diag([0.30494,0.21174,0.07092,0.05606,0.03598,0.04222])
b = [38.5390,46.1591,38.3055,40.3965,38.2704,36.3278]
alpha = np.diag([0.00838,0.00838,0.01366,0.01366,0.00922,0.00922])
beta  = [0.32767,0.32767,-0.54551,-0.54551,-0.51116,-0.51116]
d = 1200
def Pareto(xi):
```

```
   p = cvx.Variable(6)
   f_ecn = 1/2*cvx.quad_form(p,a)+b.T@p
   f_env = 1/2*cvx.quad_form(p,alpha)+beta.T@p
   fo = cvx.Minimize(xi*f_ecn+(1-xi)*f_env)
   res = [sum(p) >= d , p>=pmin, p<=pmax]
   Model = cvx.Problem(fo,res)
   Model.solve()
   return [f_ecn.value,f_env.value]
points = 10
F_ecn = np.zeros(points)
F_env = np.zeros(points)
for k in range(points):
   xi = 1/(k+1)
   F_ecn[k],F_env[k] = Pareto(xi)

plt.plot(F_ecn,F_env,marker='o')
plt.grid()
plt.xlabel('Economic')
plt.ylabel('Enviromental')
```

The reader is invited to execute the script and compare the results with previous examples.

7.3 Effect of the grid

Transmission lines impose restrictions on the economic/environmental dispatch that must be considered in the model. These constraints can be represented by three different models, namely: transportation model, linear power flow (DC-model), and non-linear power flow equations (or AC-model). The first two models are discussed in this section. The third model is studied in Chapter 10 together with the optimal power flow problem.

A first approximation of the grid constrains is based on the classic transportation model (see Example 4.5 in Chapter 4). In this model, the grid is represented by a graph $\mathcal{G} = \{\mathcal{N}, \mathcal{E}\}$ where $\mathcal{N}$ is the set of nodes and $\mathcal{E} \subseteq \mathcal{N} \times \mathcal{N}$ is the set of branches (i.e., transmission lines and transformers). Branches are included in the model by adding new variables s_j that represent the active power flow for each branch j. The main constraint is the capacity of the line/transformer:

$$|s_j| \leq s_j^{\max} \tag{7.22}$$

this constraint can be transformed into a box constraint as follows:

$$-s_j^{\max} \leq s_j \leq s_j^{\max} \tag{7.23}$$

On the other hand, loads are now distributed along the nodes forming a vector of $\mathbb{R}^n$ where n is the number of nodes. The power balance is now defined in each node as follows:

$$\sum_{i\in\Lambda_k} p_i - d_k = \sum_{j\in\Omega_k} \pm s_j \qquad (7.24)$$

where Λ_k and Ω_k are the set of generators and lines connected to node k. The sum in the right-hand side of the equation must take into account the orientation of the flux, thus s_j is positive if j departs from k and negative if it arrives to k. This concept is better understood by the following example:

Example 7.9. Consider the grid shown in Figure 7.5. Solve the economic dispatch considering the cost functions given in Table 7.1 and the transportation model of the grid. All lines have a capacity of 300 MW. The complete model is as follows:

$$\min \sum_i f_i(p_i)$$

$$d_0 + s_{01a} + s_{01b} + s_{03} + s_{04} - p_0 - p_1 = 0$$
$$d_1 - s_{01a} - s_{01b} + s_{12} + s_{15} = 0$$
$$d_2 - s_{12} + s_{25} - p_2 - p_3 = 0$$
$$d_3 - s_{03} + s_{34} - p_4 = 0 \qquad (7.25)$$
$$d_4 - s_{34} - s_{04} + s_{45} = 0$$
$$d_5 - s_{15} - s_{25} - s_{45} - p_5 = 0$$
$$-s_j^{\max} \le s_j \le s_j^{\max}, \; \forall j \in \{01a, 01b, 03, 04, 12, 15, 25, 34, 45\}$$
$$p_i^{\min} \le p_i \le p_i^{\max}, \; \forall i \in \{0, 1, 2, 3, 4, 5\}$$

Power balance equations include power demand, generation and flows in the directions shown in Figure 7.5. Two lines can connect the same nodes as is the case of $01a$ and $01b$; also, two generators can be connected to the same node as is the case of p_0, p_1 and p_2, p_3. Loads are now a vector $d = (0, 800, 0, 0, 400, 0)^\mathsf{T}$.

The script for solving this problem starts as in Example 7.2 and continues with the following code:

```
ng = 6   # number of generators
nl = 9   # number of lines
nn = 6   # number of nodes
smax = nl*[300]
d = [0,800,0,0,400,0]   # demand
Lambda = (0,0,2,2,3,5) # generators location
Omega = ((0, 1),(0,1),(0,3),(0,4),(1,2),(1,5),(2,5),(3,4),(4,5))
# grid
s = cvx.Variable(nl)   # power flows
EqB = nn*[0] # equation of balance of energy
```

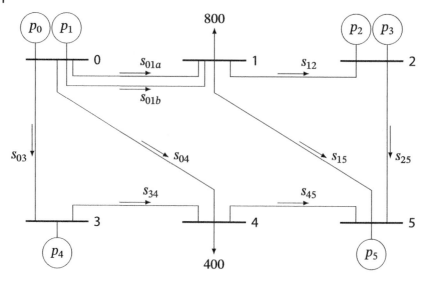

Figure 7.5 Power grid with six nodes and six generators. All lines have a capacity of 200 MW.

```
for j in range(nl):
  k = Omega[j][0]
  m = Omega[j][1]
  EqB[k] +=  s[j]    # flow in departing from k
  EqB[m] += -s[j]    # flow arriving to k
for k in range(ng):
  nl = Lambda[k]
  EqB[nl] += -p[k]
res = [p>=pmin, p<=pmax, s<=smax,  -s<=smax]
for k in range(nn):
  res += [EqB[k] + d[k] ==0]
obj = 1/2*cvx.quad_form(p,a)+b*p
Model = cvx.Problem(cvx.Minimize(obj),res)
Model.solve()
print('Generation:',np.round(p.value))
print('Flows:',np.round(s.value))
```

The active power flow in each line, including the double circuit between nodes 0 and 1, is represented by the array s; loads are represented as a vector of size six, and EqB represents Equation (7.24) i.e., the balance of energy for each node. The round solution for this model matches to Example 7.2.

Load flows are $s = (137, 137, -200, 50, -241, -284, 194, 115, -235)^{\mathsf{T}}$. Notice some power flows such as s_{03} and s_{12} are negative which indicates the power flows in the opposite direction.

The transportation model allows to include grid constraints into the economic/environmental dispatch problem. However, it is an oversimplification of the problem (it could be more useful in radial grids). We must include power flow equations into the model. A simple yet accurate approximation of the power flow equations is the linear power flow, also known as dc power flow[3]. In this case, the power flow in each branch $j = km$ is given by the following expression:

$$s_j = s_{\text{base}} \frac{\theta_k - \theta_m}{x_j} \tag{7.26}$$

where θ_k is the angle of the voltage in k and θ_m is the voltage in m; x_j is the branch impedance in per unit and s_{base} is the nominal power for per unit representation (recall p and s_j are given in MW). Let us consider the following example to understand the influence of this constraint.

Example 7.10. The economic dispatch problem presented in Example 7.9 is now solved taking into account the linear power flow. All lines have the same impedance $x = 0.02$ with $s_{\text{nom}} = 100$. The model is the same as in the previous example including additional variables and constraints as follows:

```
snom = 100
x = nl*[0.02]
th = cvx.Variable(nn)    # nodal angles
res += [th[0] == 0]       # angle reference
for j in range(nl):
  k = Omega[j][0]
  m = Omega[j][1]
  res += [x[j]*s[j] == snom*(th[k]-th[m])]
ModelLPF = cvx.Problem(cvx.Minimize(obj),res)
ModelLPF.solve()
```

The new economic dispatch was $p = (94, 100, 178, 188, 315, 325)^\top$ originated by a redistribution of the power flows which are now given by the vector $s = (133, 133, -129, 57, -300, -234, 66, 186, -157)^\top$. Notice that line 12 achieves its maximum capability. This effect was not identified with the transportation model, hence the importance of including power flow equations. The new dispatch is more expensive, but it is feasible with the conditions of the grid.

🐍

---

3 The term dc power flow comes from an analogy between the linearized model and a linear dc grid [38]. We discourage this name since it can be confused with the power flow in grids that are actually dc.

7.4 Loss equation

Losses can be included in the economic dispatch model by a simple quadratic equation and a power flow calculation. We must include both inductances and resistances of transmission lines in the model through the nodal admittance matrix or Y_{bus}. Therefore, nodal currents and nodal voltages are related by the following expression:

$$I_{bus} = Y_{bus}V_{bus} \tag{7.27}$$

We can also define the nodal impedance matrix $Z_{bus} = Y_{bus}^{-1}$; this inverse exists as long as the graph is connected, including a connection with grown, therefore, capacitance of the lines must also be included. Nodal voltages are given by Equation (7.28)

$$V_{bus} = Z_{bus}I_{bus} \tag{7.28}$$

nodal power in each node is given by the following equation

$$p_k + jq_k = v_k e^{j\theta_k} I_{bus(k)}^* \tag{7.29}$$

where v_k and θ_k are the magnitude and the angle of the voltage. Therefore, the current can be represented as function of the nodal voltages as given in Equation (7.30)

$$I_{bus(k)} = \left(\frac{p_k \cos(\theta_k) + q_k \sin(\theta_k)}{v_k} \right) + j \left(\frac{p_k \sin(\theta_k) - q_k \cos(\theta_k)}{v_k} \right) \tag{7.30}$$

On the other hand , total power losses are given by

$$p_{loss} = \text{real} \left(V_{bus}I_{bus}^H \right) \tag{7.31}$$

where $(\cdot)^H$ represents the transpose and complex conjugate. This equation can be represented as function of the real and imaginary parts of the current, namely:

$$p_{loss} = \text{real}(I_{bus}^T)R_{bus} \, \text{real}(I_{bus}) + \text{imag}(I_{bus}^T)R_{bus} \, \text{imag}(I_{bus}) \tag{7.32}$$

where $R_{bus} = \text{real}(Y_{bus}) = [r_{km}] \in \mathbb{R}^{n \times n}$. Replacing Equation (7.30) into Equation (7.32) and after a lengthy but straightforward algebraic manipulations, the following loss equation is obtained:

$$p_{loss} = p^T Bp + h^T p + w \tag{7.33}$$

where $B \in \mathbb{R}^{n \times n}$ is a positive definite matrix whose entries are given by Equation (7.34),

$$b_{km} = r_{km} \frac{\cos(\theta_{km})}{v_k v_m} \tag{7.34}$$

and h is a vector given by Equation (7.35),

$$h_k = -2 \sum_{m \in \mathcal{E}} r_{km} \frac{\sin(\theta_{km})}{v_k v_m} q_m \tag{7.35}$$

finally, w is a scalar given by Equation (7.36)

$$w = \sum_{k \in \mathcal{E}} \sum_{m \in \mathcal{E}} r_{km} \frac{\cos(\theta_{km})}{v_k v_m} q_k q_m \tag{7.36}$$

Notice that r_{km} is the input km of R_{bus} and not the resistance of the line km; in fact, it may be the case that there is no transmission line between k and m and yet, there is an r_{km} in the R_{bus} matrix. Equation (7.33) constitutes a convex quadratic form, therefore, it can be relaxed to the following convex inequality

$$p_{loss} \geq p^{\top} B p + h^{\top} p + w \tag{7.37}$$

It is common in the literature of economic dispatch to neglect h and w in the loss Equation (7.33). This approximation could be justified in cases where reactive power is negligible. However, it is advisable to consider these terms in cases where the information is available[4].

Under ideal conditions, the Lagrangian function associated to the economic dispatch with losses is given by the following expression:

$$\mathcal{L}(p, \lambda) = \sum_{k \in \mathcal{T}} f_k(p_k) + \lambda \left(d + p_{loss} - \sum_k p_k \right) \tag{7.38}$$

therefore, optimal conditions are given by

$$\frac{\partial f_k}{\partial p_k} + \lambda \left(\frac{\partial p_{loss}}{\partial p_k} - 1 \right) = 0 \tag{7.39}$$

and hence the incremental costs are

$$\lambda = \xi_k \frac{\partial f_k}{\partial p_k} \tag{7.40}$$

---

4 This simplification was also common in times where the computational resources were limited as presented in [39].

where ξ_k is a penalty factor that considers the effect of power loss, this factor is given by Equation (7.41)

$$\xi_k = \frac{1}{1 - \partial p_{\text{loss}}/\partial p_k} \tag{7.41}$$

Example 7.11. Let us solve the economic dispatch for the system presented in Example 7.2 with the following loss matrix

$$B = \begin{pmatrix} 50 & 10 & 0 & 0 & 20 & 0 \\ 10 & 50 & 0 & 0 & 0 & 10 \\ 0 & 0 & 60 & 0 & 0 & 10 \\ 0 & 0 & 0 & 350 & 20 & 0 \\ 20 & 0 & 0 & 20 & 370 & 40 \\ 0 & 10 & 10 & 0 & 40 & 480 \end{pmatrix} \times 10^{-6} \tag{7.42}$$

both h and w are zero, therefore, it is easy to transform Equation (7.37) into the following second-order constraint

$$\left\| B^{1/2} p \right\| \le z \tag{7.43}$$

where $B^{1/2}$ is the Cholesky factorization of B, and z is an auxiliar variable such that $z^2 = p_{\text{loss}}$. The code in Python is the same as in Example 7.2 modifying the set of constraints as given below:

```
B  = [[50, 10,    0,    0,   20,    0],
      [10, 50,    0,    0,    0,   10],
      [ 0,  0,   60,    0,    0,   10],
      [ 0,  0,    0,  350,   20,    0],
      [20,  0,    0,   20,  370,   40],
      [ 0, 10,   10,    0,   40,  480]]
Bchol = 1E-3*np.linalg.cholesky(B)
z   = cvx.Variable()
res = [sum(p) >= d + z**2, p>=pmin, p <= pmax]
res += [cvx.SOC(z,Bchol.T@p)]
```

The new dispatch is $p = (117, 133, 210, 225, 315, 324)^\top$, total power loss is $z^2 = 124$. Penalization factors were $\xi = (1.03, 1.02, 1.03, 1.2, 1.37, 1.52)^\top$; the first two generators had a lower penalization factor compared to the last generator, and hence it is efficient to dispatch the last generator with less power[5].

---

5 The penalization factor was calculated as `xi = 1/(1-2E-6*np.array(B)@p.value)`

7.5 Further readings

The economic/environmental dispatch has been studied for a long time in the scientific literature. It is interesting to see how the problem used to be solved on analogical computers compared to how it is solved today [40]. The algorithms can be classified into exact methods as presented in this chapter and metaheuristic algorithms as shown in [37]. Nevertheless, metaheuristic algorithms and artificial intelligence entail a lack of understanding of the problem; they are usually based on biological or social metaphors that may divert attention from the power systems problem (see [41] for an analysis about the use of these biological metaphors).

Models for the economic dispatch in actual power systems are more complex than those presented here. They can include constraints related to the stability, security, and reliability of the grid [42]. In addition, grid codes and market regulations introduce conditions that must be considered in the model. Such complex problems are usually solved by distributed algorithms [43].

Linear and piecewise linear cost functions are also standard for the economic dispatch problem. The student is invited to read [44] for a complete numerical review of different linear implementations.

7.6 Exercises

1. Plot the incremental cost function for each of the six thermal units presented in Example 7.2; plot also the optimal operation cost vs demand for the system presented in Example 7.2 with $345 \leq d \leq 1350$.
2. Formulate in Python the optimization model for Example 7.4; compare results.
3. Solve the problem presented in Example 7.2 but this time, the load varies according to to a load curve $d = (0.5, 0.3, 0.4, 0.6, 0.8, 1.0, 0.9, 0.8)^{\mathsf{T}}$. Plot the optimal incremental cost for each load.
4. Solve the multi objective economic dispatch problem presented in Example 7.8 but this time, consider the emission of NOx with $\eta = 0.0123$ and $\gamma = 0.25$ for all the units. Show the Pareto frontier as well as a plot of incremental costs vs incremental emissions.
5. Solve the problem presented in Example 7.11 without considering a limit in $p^{\max}$. Compare results considering and without considering the loss equation.
6. Write Equation (7.37) as a second-order constrain considering both h and w.
7. Power loss in a transmission line can be approximated to the following equation

$$p_{km}^{loss} \approx g_{km}\theta_{km}^2 \tag{7.44}$$

where g_{km} is the admittance of the line. Use this approximation to include loss into the economic dispatch with linearized power flow. Use the parameters of Example 7.10 with $r = 0.01$ for all transmission lines.

8. Compare results of the transportation and the linear power flow models. Experiment with different values of s^{max} and x_j.
9. Repeat the previous exercise, eliminating lines 3–4, 4–5, and 2–5.
10. Show the matrix B given in Equation (7.34) is positive semidefinite (hint: take into account that the Hadamard product of two definite matrices is also a definite matrix).

8

Unit commitment

Learning outcomes

By the end of this chapter, the student will be able to:

- Identify the difference between the economic dispatch and the unit commitment
- Solve basic problems of deterministic unit commitment
- Include transmission constraints into the model by a linear power flow formulation.

8.1 Problem definition

The economic dispatch presented in Chapter 7 is a continuous problem decoupled in time; the latter means that current decisions do not affect future operation. However, this picture is incomplete since a power system presents a dynamic behavior due to loads' daily changes. Therefore, a dynamical model is required.

The starting-up of a thermal power plant is not instantaneous since the boiler requires suitable pressure and temperature conditions to generate power, as shown in Figure 8.1. Shutting-down is not instantaneous either. Besides, several physical and economic limitations such as the minimum operative time and the maximum off-line time must be considered in the model. All these constraints introduce binary variables into the optimization problem. Thus, the unit starts generating power when the temperature is between T_{min} and T_{nom}; below T_{min} the unit requires fuel to maintain the temperature, but the output power is zero. This implies costs that must be included in an optimization model named unit commitment. This problem is discrete since the

Mathematical Programming for Power Systems Operation: From Theory to Applications in Python. First Edition. Alejandro Garcés.
© 2022 by The Institute of Electrical and Electronics Engineers, Inc. Published 2022 by John Wiley & Sons, Inc.

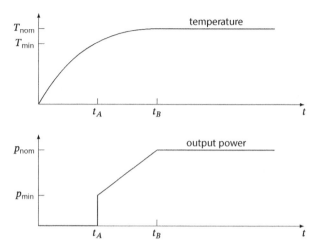

Figure 8.1 Simplified model for the starting-up of a thermal power plant.

time in which the unit is connected (committed) or disconnected from the grid (de-committed) is part of the decision; hence the problem dynamic. Additional constraints are also included in the model, related to the ramps of start-up and start-down in each thermal unit. In the following sections, we present the basic model with the most common constraints. As in all the previous chapters, we offer only toy-models in order to understand the problem.

8.2 Basic unit commitment model

Unlike the economic dispatch, the unit commitment requires considering each individual unit into the model. For example, a combined cycle power plant may have five gas units and two steam. Each of these units must be considered in the model since they have a different dynamic performance. Gas units are more flexible with relatively fast start-up time compared to steam units.

We consider a time horizon $T = \{0, 1, \dots, T\}$, for example one day or one week with discrete steps in hours. Thermal units are grouped in a set $\mathcal{T}$ with three types of costs, namely:

$$f_{\text{operation}} + f_{\text{start-up}} + f_{\text{shut-down}} \tag{8.1}$$

Operation costs ($f_{\text{operation}}$) are linear or quadratic functions as presented in Chapter 7; start-up ($f_{\text{start-up}}$) and shut-down costs ($f_{\text{shut-down}}$) are linear functions that depend of the status (on/off) of each power unit. Therefore, binary variables are required to identify each transition. A set of variables ζ_{kt} are defined such that $\zeta_{kt} = 1$ if the unit k is operating at the time t. The starting-up action is defined by another binary variable μ_{kt} such that $\mu_{kt} = 1$ if the unit k was disconnected in time $t - 1$ and connected in the time t. Similarly, a binary variable δ_{kt} is defined such that $\delta_{kt} = 1$ if the unit was connected at time $t - 1$ but disconnected at time t. Additional constraints are required in order to identify starting-up and shutting-down as presented below:

$$\mu_{kt} - \delta_{kt} = \zeta_{kt} - \zeta_{kt-1} \tag{8.2}$$

$$\mu_{kt} + \delta_{kt} \leq 1 \tag{8.3}$$

$$\zeta_{kt}, \mu_{kt}, \delta_{kt} \in \{0, 1\} \tag{8.4}$$

Notice that these constraints uniquely meet the conditions presented in Table 8.1 for binary variables ζ, μ, δ. Thus, $\mu = 1$ only in the case the unit starts to operate; similarly, $\delta = 1$ only in the case the unit is disconnected.

Table 8.1 Logic table for operation (ζ), start-up (μ) and shut-down (δ) conditions.

ζ_{kt-1}	ζ_{kt}	μ_{kt}	δ_{kt}
0	0	0	0
0	1	1	0
1	0	0	1
1	1	0	0

With these binary variables, it is possible to define the cost functions for a time horizon, as presented below:

$$f_{\text{operation}} = \sum_{k \in \mathcal{J}} \sum_{t \in T} a_{kt} p_{kt}^2 + b_k p_{kt} + c_k \zeta_{kt} \tag{8.5}$$

$$f_{\text{start-up}} = \sum_{k \in \mathcal{J}} \sum_{t \in T} c_k^{\text{up}} \mu_{kt} \tag{8.6}$$

$$f_{\text{shut-down}} = \sum_{k \in \mathcal{J}} \sum_{t \in T} c_k^{\text{down}} \delta_{kt} \tag{8.7}$$

These functions may be more complex in practice. For example, the start-up cost depends on how long the unit was de-committed since the cost is lower as the initial temperature is higher. A detail model of the start-up cost may include exponential cost functions as presented in [45]. However, it is common practice to linearize these functions. Notice that ζ_k affects the fixed costs in $f_{\text{operation}}$ under the assumption that the unit incurs this cost only when connected.

The variable ζ_k affects also the operation limits of the thermal units as follows:

$$\zeta_{kt} p_k^{\min} \leq p_{kt} \leq p_k^{\max} \zeta_{kt} \tag{8.8}$$

thus, $p_{kt} = 0$ when $\zeta_{kt} = 0$. The model is completed with the power balance at each time t as presented below:

$$\min \sum_{k \in \mathcal{T}} \sum_{t \in \mathsf{T}} a_{kt} p_{kt}^2 + b_k p_{kt} + c_k \zeta_{kt} + c_k^{\text{up}} \mu_{kt} + c_k^{\text{down}} \delta_{kt}$$

$$\sum_{k \in \mathcal{T}} p_{kt} = d_t, \ \forall t \in \mathsf{T}$$

$$\zeta_{kt} p_k^{\min} \leq p_{kt} \leq p_k^{\max} \zeta_{kt}, \ \forall t \in \mathsf{T}, k \in \mathcal{T} \tag{8.9}$$

$$\mu_{kt} - \delta_{kt} = \zeta_{kt} - \zeta_{kt-1}, \ \forall t \in \mathsf{T}, k \in \mathcal{T}$$

$$\mu_{kt} + \delta_{kt} \leq 1, \ \forall t \in \mathsf{T}, k \in \mathcal{T}$$

$$\zeta_{kt}, \mu_{kt}, \delta_{kt} \in \{0,1\}, \ \forall t \in \mathsf{T}, k \in \mathcal{T}$$

This is the basic model of the unit commitment problem. However, it can be complemented with additional constraints related to the grid. Practical applications combine the hydrothermal schedule with the unit commitment and even with the ac optimal power flow. Therefore, Equation (8.9) may be considered as a toy-model used to understand the problem. Let us see the model in practice:

Example 8.1. Let us solve the basic unit commitment problem for a system with three thermal units. Parameters of the system are presented directly in the following Python script:

```
a = np.array([0.0004984, 0.001246, 0.00623 ])
b = np.array([16.821 , 40.6196, 21.9296])
c = np.array([220.4174, 161.8554, 171.2004])
c_up = np.array([124.69, 249.22, 0])
z_ini = np.array([1,1,0])
pmax = np.array([220, 100, 20])
pmin = np.array([100,10,10])
d = np.array([178.690,168.450,161.840,157.830,158.160,163.690,
              176.860,194.210,209.670,221.540,233.180,240.820,
              247.030,248.470,253.830,260.900,261.120,251.680,
              250.890,242.100,242.050,231.680,205.070,200.690])
```

These values were adapted from [46] for 24h operation with $c^{\text{down}} = 0$. The optimization model is easily translated from Equation (8.9) to a Python script, with $\mathsf{T} = \dim(\mathsf{T})$ and $n = \mathcal{T}$, as presented below:

```
T = len(d)
n = len(a)
zeta  = cvx.Variable((n,T), boolean=True)
mu    = cvx.Variable((n,T), boolean=True)
delta = cvx.Variable((n,T), boolean=True)
p     = cvx.Variable((n,T))

fop = 0   # operation cost
fsup = 0  # start-up cost
res = []
for t in range(T):
    for k in range(n):
        fop = fop + a[k]*p[k,t]**2+b[k]*p[k,t]+c[k]*zeta[k,t]
        fsup = fsup + c_up[k]*mu[k,t]
        res += [p[k,t] >= pmin[k]*zeta[k,t]]
        res += [p[k,t] <= pmax[k]*zeta[k,t]]

for t in range(T):
    res += [cvx.sum(p[:,t])==d[t]]

for t in range(1,T):
    for k in range(n):
        res += [mu[k,t]-delta[k,t] == zeta[k,t]-zeta[k,t-1]]
        res += [mu[k,t]+delta[k,t] <= 1]

for k in range(n):
    res += [mu[k,0]-delta[k,0] == zeta[k,0]-z_ini[k]]
    res += [mu[k,0]+delta[k,0] <= 1]

obj = cvx.Minimize(fop+fsup)
UnitC = cvx.Problem(obj,res)
UnitC.solve()
print(UnitC.status, obj.value)
```

The optimal value was 100807. Binary variables can be plotted as follows:

```
plt.subplot(4,1,1)
plt.plot(p.value.T)
plt.subplot(4,1,2)
plt.pcolor(zeta.value)
plt.subplot(4,1,3)
plt.pcolor(mu.value)
plt.subplot(4,1,4)
plt.pcolor(delta.value)
plt.show()
```

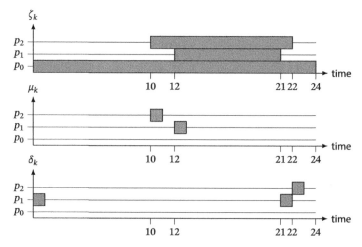

Figure 8.2 Unit commitment for a system with three thermal units.

Figure 8.2 shows the results for the binary variables of the problem. Notice that μ and δ properly identify committed and de-committed time of each thermal units.

8.3 Additional constraints

The unit commitment problem can be complemented with additional constraints. For instance, a thermal unit may require maintenance at a particular time. In that case, binary variables ζ_{kt} must be zero during the time expected to run maintenance.

Likewise, thermal units have minimum uptime and downtime. The former is the minimum time the unit must be committed once it is turned on, while the latter is the minimum time the unit is de-committed before it can be turned on again. For example, if the minimum up time is 4h, we can define the following constraint:

$$\zeta_{k+1} + \zeta_{k+2} + \zeta_{k+3} + \zeta_{k+4} \geq 4\mu_k \tag{8.10}$$

This constraint force $\zeta_{k+1} = \cdots = \zeta_{k+4} = 1$ when $\mu_k = 1$, that is to say, when the unit is turned on. The same can be done for the minimum downtime. All these constraints are affine, and hence, the model remains tractable.

Thermal units cannot achieve full load instantaneously; likewise, they cannot pass instantaneously from full to zero load; therefore, the turning-on and turning-off process require to be gradual. This is represented by ramping limits as follows:

$$p_{kt} - p_{kt-1} \leq \rho_k^{up} \tag{8.11}$$

$$p_{kt-1} - p_{kt} \leq \rho_k^{down} \tag{8.12}$$

On the other hand, it is required to guarantee a fixed amount of power to hedge the system against the sudden changes of load and generation. This quantity, known as spinning reserve σ, is included in the model as presented below:

$$\sum_{k \in \mathcal{T}} \zeta_{kt} p_k^{max} - p_{kt} \geq \sigma_t, \ \forall t \in \mathsf{T} \tag{8.13}$$

Notice the spinning reserve may be different for each time. For example, it could be higher for the periods of peak load where load changes are greater.

Thermal units may have physical operation limitations that create prohibited operating zones of power. These restricted zones must be included in the model as additional binary variables and/or constraints.

8.4 Effect of the grid

Just like the economic dispatch and the hydrothermal schedule, the unit commitment may include power flow constraints, either as dc or ac formulations. For easy explanation, we present here the transportation model. AC power flow equations can be included with linear or conic approximations, as explained in Chapter 10.

Example 8.2. Consider the problem presented in Example 8.1 with the grid shown in Figure 8.3. All lines have a maximum capacity of 100MW.

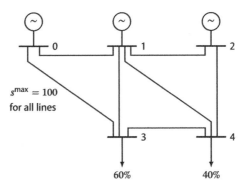

Figure 8.3 Grid constraints for the unit comment problem.

Let us code this grid in Python but this time, we use the module NetworkX that allows to operate graphs; the command DiGraph generate an oriented graph with the connections depicted in Figure 8.3. The graph can be plotted as presented below:

```python
import networkx as nx
grid = nx.DiGraph([(0,1),(0,3),(1,2),(1,3),(1,4),(2,4),(3,4)])
smax = np.array([100,100,100,100,100,100,100]) # smax lines
smax = smax;
nl = 7 # number of lines
nn = 5 # number of nodes
plt.figure()
nx.draw(grid,with_labels=True)
plt.show()
```

We neglect power loss and therefore, nodal power can be represented as function of the power flows, as follows:

$$p^{node} = Ap^{flow} \tag{8.14}$$

where A is the incidence matrix which is calculated by the module NetworkX:

```python
A = nx.incidence_matrix(grid,oriented=True)
```

Finally, we define a new variable p^{flow} that represents the power flow in each line and each time. We assume thermal units are connected to nodes 0 to 2 and load is distributed between nodes 3 and 4 with 60% in node 3 and 40% in node 4. These considerations are transformed in constrains as presented below:

```python
     = cvx.Variable((nl,T))    # power flows
p_node = cvx.Variable((nn,T))    # nodal powers
for t in range(T):
    res +=[ p_flow[:,t] >= -smax]
    res +=[ p_flow[:,t] <= smax]
    res +=[ p_node[:,t] == A@p_flow[:,t]]
    res +=[ p_node[0,t] == p[0,t]]
    res +=[ p_node[1,t] == p[1,t]]
    res +=[ p_node[2,t] == p[2,t]]
    res +=[ p_node[3,t] == -d[t]*0.6]
    res +=[ p_node[4,t] == -d[t]*0.4]

obj = cvx.Minimize(fop)
UnitG = cvx.Problem(obj,res)
UnitG.solve()
print(UnitG.status, obj.value)
```

The optimal point is now 106795. As expected, the unit commitment was affected by grid constraints. However, in this case, the starting-up and shutting-down conditions are the same as in Example 8.1. This is not the general case.

However, this system is tiny compared to current power systems, and the grid has enough capability to transport all generated power. The reader is invited to experiment with this model, modifying $s^{\max}$ and adding practical constraints in the model.

8.5 Further readings

A complete review of the unit commitment problem can be found in [47] that includes stochastic and robust versions of the problem. More details about the mathematical formulation and especially the use of binary variables can be found in [45]. The model may include ac power flow constraints as given in [46], where different test systems are available. An extension of the problem for power distribution systems with renewable energy and storage devices can be studied in [48].

Besides the mixed-integer programming approach presented in this chapter, the unit commitment problem can be solved by heuristic techniques and dynamic programming. A complete review of these approaches can be found in [49].

8.6 Exercises

1. Make a comparative table between the unit commitment problem and the economic dispatch.
2. Solve the problem presented in Example Example 8.1 as an economic dispatch problem, that is to say, without including start-up and shut-down costs. Compare the results with the unit commitment.
3. Solve the unit commitment problem in the system presented in Example 8.1 considering a spinning reserve of $\sigma = 20\ MW$. Compare results.
4. Include now ramping limits as $\rho^{\text{up}} = \rho^{\text{down}} = (55, 50, 20)^{\top}$.
5. Include a minimum up a limit of 4h for all thermal units.
6. Include power loss into the unit commitment model. Use a quadratic approximation as presented in Example 7.11. Compare results.
7. Solve the unit commitment problem considering the transportation model (Example 8.2) without using the module NetworkX.
8. Solve the problem presented in Example 8.2 using a linear power flow instead of the transportation model. Use $x_{km} = 0.01\text{pu}$. Compare results.

9. Solve the problem presented in Example 8.2, but this time the load is shared 50% between nodes 3 and 4.

10. Formulate the unit commitment problem considering the ac power flow constraints. Identify the main characteristics of this model (we will learn how to solve this type of problems in Chapter 10)

9

Hydrothermal scheduling

Learning outcomes

By the end of this chapter, the student will be able to:

- Formulate the problem of the hydrothermal dispatch.
- Include hydraulic chains into the model.
- Study some non-linear constraints related to the model of hydraulic units.

9.1   Short-term hydrothermal coordination

Hydropower is a renewable energy source with high potential around the world [50]. Despite its advantages, such as high flexibility and fast dynamic response, hydropower generation is highly vulnerable to complex weather patterns such as El Niño-southern oscillation. Therefore, systems with high hydropower generation capability are usually complemented with thermal units, and hence economic dispatch requires considering both hydroelectric and thermal power stations. This problem is more complex than the economic dispatch in all-thermal units for two reasons: first, the problem is coupled in the time; and second, the system may have hydraulic chains. The first aspect implies that an operation decision at one time can affect the future operation of the grid. The second aspect implies that an operative decision in a hydroelectric unit upstream in a river (or hydraulic chain) may affect the hydroelectric units placed downstream in the same hydraulic chain. These two aspects are studied in this chapter.

Hydrothermal scheduling requires considering the dynamics of the electric part and the dynamics of the hydraulic system, which includes the change in

Mathematical Programming for Power Systems Operation: From Theory to Applications in Python. First Edition. Alejandro Garcés.
© 2022 by The Institute of Electrical and Electronics Engineers, Inc. Published 2022 by John Wiley & Sons, Inc.

the volume of the reservoirs and water discharges and spillage. These variables may be related to other uses of the reservoir, for example, irrigation; hence, additional constraints must be included in the model. These constraints can also be related to the safety limits of the volume and/or discharges of the reservoir.

The level of detail of the model of the hydroelectric system may vary from one system to another. There are many types of hydroelectric and reservoirs, and each one has a different type of model. Thus, the mathematical relation between volume/water discharge and power may be linear or non-linear. In addition, hydraulic chains can introduce delays in the inflows that affect the entire system's dynamic. To do this, we must add non-linear constraints related to losses and cost of thermal units, and obtaining a model highly non-linear that requires to be solved in real-time.

On the other hand, the modern power system may include pumped hydro-electric storage power plants. This type of storage is becoming more popular with the high penetration of renewable energies. Pump energy storage is just hydroelectric units with two reservoirs and a reversible capability. Water can be pumped from the low reservoir to an upper reservoir to store energy. The effect of other renewable energies, such as wind and solar, must be included in the model as well.

9.2 Basic hydrothermal coordination

Let us consider a hydrothermal system where the units are grouped in two sets, $\mathcal{H}$ for hydraulic units and $\mathcal{T}$ for thermal units. Incremental costs of hydroelectric units are usually neglected in practice, then the objective function consists of minimizing costs of thermal units, just as in the conventional economic dispatch. However, the optimization model must include physical constraints related to hydroelectric power plants. The problem becomes coupled in the time since an operative decision in one instant can affect the subsequent operation. Moreover, decision variables are not only the generated power but also the water discharge, spillage, and volume of the reservoirs.

Generated power in each hydroelectric unit $i \in \mathcal{H}$ can be calculated as given in Equation (9.1),

$$p_i = (\rho g \eta_i h_i) q_i \tag{9.1}$$

Where ρ is the water density ($\approx 100 \text{kg/m}^3$); g is the acceleration of the gravity 9.81m/s^2; h_i is the head of reservoir, measured in meters; η_i is the efficiency of the group turbine-generator; and q_i is the water discharge in m^3/s (i.e., the flow passing through the turbine). The three most common types of turbines are Pelton, Francis, and Kaplan. Pelton is impulse turbines used for high-head plants, while Francis and Kaplan are reaction turbines used for medium and low heads, respectively. In regards to the water flow, Pelton turbines are used for relatively low water flow rates while Francis and Kaplan are used for high water flow [51]. The head and efficiency of the unit depending on the volume of the reservoir and the water discharge; however, they can be considered constant for hydro plants with large reservoir capability and in cases where the powerhouse is placed at a long distance below the dam. In those cases, generated power can be considered proportional to the water discharge, as given in Equation (9.2):

$$p_i = \mu_i q_i, \ \forall i \in \mathcal{H} \tag{9.2}$$

where $\mu_i = \rho g \eta_i h_i$ is called turbine factor.

The dynamics of the reservoir must be considered into the model. It includes the volume of the reservoir v, water discharge q, inflows a, and spillage s as given in Equation (9.3),

$$v_{it+1} = v_{it} + \Delta T(a_{it} - q_{it} - s_{it}), \ \forall i \in \mathcal{H}, t \in \mathsf{T} \tag{9.3}$$

where ΔT is the discretization of the time in the operation horizon $\mathsf{T} = \{0, 1, \dots, T\}$ (usually $\Delta T = 1h$). The horizon may be one day, one week, or one month, discretized in hours or even minutes, according to the desired level of detail. In this model, we assume an accurate forecasting of the inflows, so the model is deterministic.

The entire model for the short-term hydrothermal dispatch is presented below:

$$\min \sum_{t \in \mathsf{T}} \sum_{k \in \mathcal{J}} f_k(p_{kt})$$

$$p_{it} = \mu_i q_{it}, \qquad\qquad \forall i \in \mathcal{H}$$

$$v_{it+1} = v_{it} + a_{it} - q_{it} - s_{it}, \qquad\qquad \forall i \in \mathcal{H}, t \in \mathsf{T}$$

$$\sum_{i \in \mathcal{H}} p_{it} + \sum_{k \in \mathcal{J}} p_{kt} = d_t, \qquad\qquad \forall t \in \mathsf{T}$$

$$v_i^{\min} \leq v_{it} \leq v_i^{\max} \qquad\qquad \forall i \in \mathcal{H}, t \in \mathsf{T}$$

$$q_i^{\min} \leq q_{it} \leq q_i^{\max} \qquad\qquad \forall i \in \mathcal{H}, t \in \mathsf{T} \tag{9.4}$$

$$p_i^{\min} \leq p_{it} \leq p_i^{\max} \qquad\qquad \forall i \in \mathcal{H}, t \in \mathsf{T}$$

$$p_k^{\min} \leq p_{kt} \leq p_k^{\max} \qquad\qquad \forall k \in \mathcal{J}, t \in \mathsf{T}$$

$$0 \leq s_{it} \leq s_i^{\max} \qquad\qquad \forall i \in \mathcal{H}, t \in \mathsf{T}$$

$$v_{i0} = v_i^{\text{initial}}$$

$$v_{iT} = v_i^{\text{final}}$$

The model is similar to the conventional economic dispatch of thermal units; just, in this case, the power balance equation includes both thermal and hydropower plants. The objective function is associated with thermal power plants' operation cost since the incremental operating cost of hydroelectric units is almost zero (a suitable approximation in practice). The objective function can be either quadratic or linear, according to the cost model of thermal units. The rest of the constraints are related to the dynamics of the reservoir, generated power of the hydroelectric, and box constraints that represent the limits of each variable. Most of the constraints are affine in this basic model, and hence the problem is convex, although it may present a high number of decision variables. Initial and final values of the volume in the reservoirs are obtained by a medium or long-term model of hydrothermal coordination[1].

Example 9.1. Let us consider a hydrothermal system with one hydropower plant p_H and one thermal unit p_T with a linear cost function $f = 19.2p_T$. Other parameters of the system are $p_T^{\min} = 0$, $p_T^{\max} = 250$, $p_H^{\min} = 0$, $p_H^{\max} = 150$, $\mu_H = 8.5$, $v^{\max} = 150$, $v^{\min} = 80$. The initial volume of the reservoir is $v^{\text{initial}} = 150$ while the final volume is required to be at least $v^{\text{final}} = 80$. The demand and inflows are presented in Figure 9.1.

The code in Python is a direct representation of Equation (9.4) as presented below:

```
import numpy as np
import cvxpy as cvx
import matplotlib.pyplot as plt
d = (137, 139, 136, 129, 129, 141, 165, 200, 224, 232, 223, 231,
     223, 220, 213, 207, 213, 214, 224, 228, 224, 212, 185, 159)
         #load
a = (10,9,8,7,7,7,8,9,10,10,10,9,8,8,8,8,8,9,9,9,9,9,9,9)
     # inflows
pH = cvx.Variable(24)   # power hydro unit
pT = cvx.Variable(24)   # power thermo unit
q  = cvx.Variable(24)   # water discharge
s  = cvx.Variable(24)   # spillage
v  = cvx.Variable(25)   # volume
cost = 0
res = [v[0]==150, v[24]==80]
for t in range(24):
  cost = cost + 19.2*pT[t]
  res += [pH[t] == 8.5*q[t]]
  res += [pH[t] + pT[t] == d[t]]
  res += [80 <= v[t], v[t] <= 150]
```

---

1 Long-term hydrothermal coordination is a stochastic problem related to operational planning. Although its model is similar to the short-term hydrothermal coordination, the study of this problem is beyond the objectives of this book. Interested reader is invited to see [52].

```
res += [0 <= pT[t], pT[t] <= 250]
res += [0 <= pH[t], pH[t] <= 150]
res += [s[t] >= 0]
res += [v[t+1] == v[t] + a[t] - s[t] - q[t]]
HydTh = cvx.Problem(cvx.Minimize(cost), res)
HydTh.solve()
```

Despite being a system with only two units, the model presents 121 decision variables, a very high number compared to a dispatch in an all-thermal system. Nevertheless, the model is linear or, at most quadratic. Generated power in each unit is given in Figure 9.2 together with the total demand. The power generated by the hydroelectric unit is very flat, following a curve that guarantees the reservoir's initial and final volume and minimizes the use of the thermal unit. The power in the thermal unit tries to follow the demand at a minimum cost.

9.3 Non-linear models

The linear model given by Equation (9.2) may be insufficient to represent accurately hydroelectric units, specially in the case of variable-head hydro plants. In those cases, a quadratic model that relates the output power with the water discharge and the volume of the reservoir is required as given in Equation (9.5),

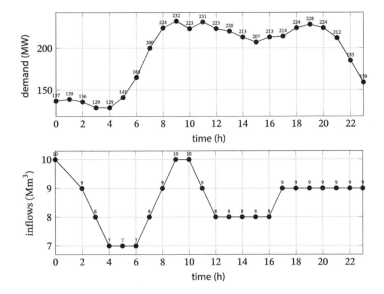

Figure 9.1 Power demand and inflows for a hydrothermal system.

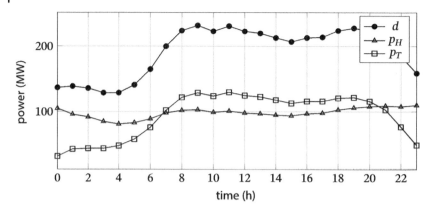

Figure 9.2 Power demand and generated power.

$$p_{it} = -x^\mathsf{T} A x + b^\mathsf{T} x + c \tag{9.5}$$

where $x = (v_{it}, q_{it})^\mathsf{T}$ and A is a real-square 2×2 matrix; this equation includes the effect of the turbine efficiency as well as the head variations that in most hydroturbines, is given by the so called hill-chart curve [53]. This curve describes a concave surface and hence A can be adjusted to a positive definite matrix making Equation (9.5) a concave quadratic form[2] as shown in Figure 9.3.

Equation (9.5) is clearly non-convex, however, it can be approximated either to a semidefinite or a second-order constraint as presented in [54] and [55], respectively.

On the other hand, the effect of the grid can be considered into the model, just as in the case of thermal units, as presented in the following example adapted from [56].

Example 9.2. A hydrothermal system consists on four units labeled as $\{0, 1, 2, 3\}$ where $\{0, 1\}$ are hydroelectric units and $\{2, 3\}$ are thermal power plants. Each hydroelectric unit has a non-linear relation to the water discharge given by the following concave quadratic functions:

$$p_0 = -0.58q_0^2 + 3.60q_0 + 0.89 \tag{9.6}$$

$$p_1 = -0.78q_1^2 + 3.96q_1 + 1.13 \tag{9.7}$$

---

2 Notice the sign minus in Equation (9.5).

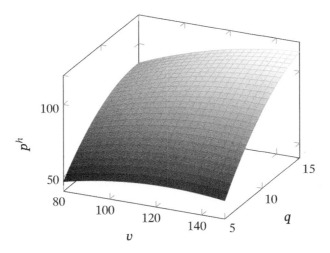

Figure 9.3 Quadratic function for a hydropower unit.

The minimum permissible water discharge is $q = 0.5$ for both units. Generation costs of the thermal plants are adjusted to the following quadratic forms:

$$f_2(p_2) = 0.80p_2 + 0.02p_2^2 \tag{9.8}$$

$$f_3(p_3) = 0.78p_3 + 0.03p_3^2 \tag{9.9}$$

In addition, power losses are given by the following loss-formula matrix

$$B = \begin{pmatrix} 0.05 & -0.02 & 0.01 & 0.00 \\ -0.02 & 0.06 & -0.02 & 0.01 \\ 0.01 & -0.02 & 0.04 & 0.00 \\ 0.00 & 0.01 & 0.00 & 0.02 \end{pmatrix} \tag{9.10}$$

Initial and final volume are 10 pu for reservoir 0 and 12 pu for reservoir 1. Inflows and power demand for 12 h operation are included in the following script:

```
import cvxpy as cvx
p  = cvx.Variable((12,4))   # hydro = {0,1} thermal = {2,3}
pL = cvx.Variable(12)       # losses
v  = cvx.Variable((13,2))   # volumne
q  = cvx.Variable((12,2))   # water discharge
a  = [[0, 0.6, 1.2, 1.2, 1.2, 1.8, 2.4, 1.5, 1.2, 0.9, 0, 0],
      [0, 0.0, 0.0, 1.5, 3.0, 4.5, 4.5, 1.5, 0.0, 0.0, 0, 0]]
      # inflows
B  = [[ 0.05,-0.02, 0.01, 0.00],
      [-0.02, 0.06,-0.02, 0.01],
```

```
      [ 0.01,-0.02,  0.04,  0.00],
      [ 0.00,  0.01,  0.00,  0.02]]   # loss matrix
d = [8,  7,  7,  6,  7,  8,  9,  8,  7,  7,  8,  8]   # demand
cost = 0
res = [v[0,0] == 10,  v[0,1] == 12,  v[12,0]==10,  v[12,1] == 12]
for t in range(12):
  cost = cost + 0.02*p[t,2]**2 + 0.80*p[t,2]
  cost = cost + 0.03*p[t,3]**2 + 0.78*p[t,3]
  res += [sum(p[t,:]) == d[t] + pL[t]]
  res += [pL[t] >= cvx.quad_form(p[t,:],B)]
  res += [q[t,:] >= 0.5]
  res += [p[t,:] >= 0.0]
  res += [p[t,0] + 0.58*q[t,0]**2 - 3.60*q[t,0] + 0.89 <= 0 ]
  res += [p[t,1] + 0.78*q[t,1]**2 - 3.96*q[t,1] + 1.13 <= 0 ]
  res += [v[t+1,0] == v[t,0] + a[0][t] - q[t,0]]
  res += [v[t+1,1] == v[t,1] + a[1][t] - q[t,1]]
HydTh = cvx.Problem(cvx.Minimize(cost), res)
HydTh.solve()
```

In this case, the code was a little different from the previous example. First, generation power was saved in a single vector where the two first terms correspond to hydraulic units and the last two to thermal units. Spillages were set to zero, and quadratic functions were included as inequality constraints. Notice this is an approximation that requires to be evaluated after solving the optimization problem. As always, the reader is invited to execute and experiment with the script.

9.4 Hydraulic chains

Generation systems may contain hydraulic chains where the spillage and water discharge of one unit are part of the inflows of other units downstream, as shown in Figure 9.4. Hydraulic chains constitute a hydraulic network where each node represents a reservoir. Therefore, a balance of flows must be included for each of these reservoirs as given in Equation (9.11),

$$v_{it} = v_{it-1} + a_{it} - q_{it} - s_{it} + \sum_{j \in \Omega_i}(q_{jt-\tau} + s_{jt-\tau}) \; \forall i \in \mathcal{H}, t \in \mathsf{T} \qquad (9.11)$$

where Ω_i represents the set of nodes that are connected to reservoir i in the hydraulic chain[3]. Notice that Equation (9.11) is an affine constraint, hence the problem remains convex. On the other hand, flows that come from an upper reservoir to a lower reservoir do not arrive immediately. Therefore, a time delay

---

3 For the hydraulic chain given in Figure 9.4, we have that $\Omega_2 = \{0, 1\}$ and $\Omega_3 = \{2\}$, whereas Ω_0 and Ω_1 are just empty.

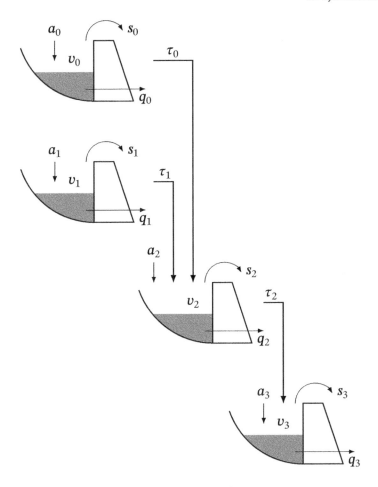

Figure 9.4 Example of a hydraulic chain where a_i represents inflows, v_i volume, q_i flow, s_i spillage, and τ_i delays.

τ_i must be considered in each branch of the hydraulic network. This constrain is not relevant for medium and long-term hydrothermal coordination, but it is important in short-term scheduling. Box constraints related to spillage and water discharge must be carefully considered in hydraulic chains. Safety levels of the rivers and other uses of the reservoir introduce additional constraints to the problem [57].

Example 9.3. A power generation system consists on a large thermal unit and the hydraulic chain depicted in Figure 9.4. Operation cost of the thermal unit is linear and given by $f(p) = 19.2p$. Parameters of the system including demand and inflows are coded in Python as follows:

```python
numh = 4
vmin = [80,60,100,70]
vmax = [150,120,240,160]
vini = [100,80,170,120]
vend = [120,80,170,100]
qmin = numh*[0]
qmax = [15,15,30,30]
smin = numh*[0]
smax = numh*[10]
pHmax = numh*[500]
pHmin = numh*[0]
pTmax = 2500
pTmin = 0
a = [[ 9.0,   7.5,    2.5,    2.8],
     [ 9.2,   8.0,    3.0,    2.4],
     [ 9.5,   8.8,    4.0,    1.6],
     [ 9.6,   9.0,    4.5,    1.0],
     [ 9.8,   9.3,    4.3,    1.0],
     [ 9.9,   9.5,    4.0,    1.0],
     [10.0,  10.0,    3.0,    1.0],
     [10.3,  10.2,    2.0,    1.3],
     [10.5,  10.3,    1.5,    1.5],
     [11.0,  10.3,    1.0,    1.6],
     [11.2,  10.5,    1.0,    1.7],
     [11.5,  10.4,    1.8,    1.5],
     [11.4,  10.3,    2.3,    1.5],
     [11.3,  10.0,    3.0,    1.3],
     [11.2,   9.8,    3.0,    1.2],
     [10.0,   9.5,    2.8,    1.2],
     [ 9.3,   9.3,    2.5,    1.2],
     [ 8.6,   9.0,    2.0,    1.0],
     [ 7.5,   8.8,    1.8,    1.0],
     [ 7.0,   8.7,    1.6,    0.8],
     [ 7.2,   8.6,    1.6,    0.8],
     [ 7.3,   8.3,    1.8,    0.8],
     [ 7.4,   8.0,    2.0,    0.8],
     [ 7.5,   8.0,    2.0,    0.8]]
d = [685,695,680,645,645,705,825,1000, 1120,1160,1115,1155,
     1115,1100,1065,1035,1065,1070,1120,1140, 1120,1060,925,795]
miu = [6.5,5.5,9.4,4.7]
```

A linear model for the hydrothermal scheduling is presented below, where delays were neglected:

```python
import cvxpy as cvx
pT = cvx.Variable(24)
pH = cvx.Variable((24,numh))
v  = cvx.Variable((25,numh))
q  = cvx.Variable((24,numh))
s  = cvx.Variable((24,numh))
cost = 5000
res = []
for t in range(24):
  cost=cost +  19.2*pT[t]
  res+=[pT[t] >= pTmin, pT[t] <= pTmax]
  res+=[sum(pH[t,:])+pT[t]==d[t]]
  res+=[v[t+1,0]==v[t,0]+a[t][0]-s[t,0]-q[t,0]]
  res+=[v[t+1,1]==v[t,1]+a[t][1]-s[t,1]-q[t,1]]
  res+=[v[t+1,2]==v[t,2]+a[t][2]-s[t,2]-q[t,2]
        +s[t,0]+q[t,0]+s[t,1]+q[t,1]]
  res+=[v[t+1,3]==v[t,3]+a[t][3]-s[t,3]-q[t,3]+s[t,2]+q[t,2]]
  for k in range(numh):
    res += [v[0,k] == vini[k], v[24,k] == vend[k]]
    res += [v[t,k] >= vmin[k], v[t,k] <= vmax[k]]
    res += [q[t,k] >= qmin[k], q[t,k] <= qmax[k]]
    res += [s[t,k] >= smin[k], s[t,k] <= smax[k]]
    res += [pH[t,k] >= pHmin[k], pH[t,k] <= pHmax[k]]
    res += [pH[t,k] == miu[k]*q[t,k]]
HydTh = cvx.Problem(cvx.Minimize(cost), res)
HydTh.solve()
```

The model does not grow significantly in the number of variables. It is required only to include additional constraints related to the balance of flows in each reservoir in the hydraulic chain. Otherwise, the model is the same as the previous cases. Spillages are important in hydraulic chains since they affect the power production of hydraulic units placed downstream. In some cases, the optimization model could introduce spillages in one reservoir in order to supply another reservoir and increase power generation.

9.5 Pumped hydroelectric storage

Pumped hydroelectric storage is a classic technology that is recovering attention due to the increasing penetration of wind and solar generation. These new types of renewable resources present high variability, and hence energy storage is required. A pumped energy storage system consists of two reservoirs connected with a combined pump/turbine system as shown in Figure 9.5. In generation mode, water flows from the upper to the lower reservoir, generating

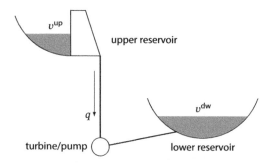

Figure 9.5 Schematic representation of a pumped hydroelectric storage system.

power to the electric grid. In charging mode, water is pumped from the lower to the upper reservoir taking electric power from the grid.

Finding a suitable place for building a pumped hydroelectric storage system is the main limitation of this technology. However, construction of a lower reservoir placed deep underground and directly below the upper reservoir can reduce this limitation [58]. Efficiency and energy density is another main limitation; the total efficiency of the existing pumped hydroelectric storage system is around 70–85% [59]. However, the use of variable speed systems can increase these values [60].

Compared to other storage technologies, pumped hydroelectric have the largest capacity in both energy and power, which varies from 1 to 300 MW. The turbine/pump system is usually placed just below the upper reservoir, connected with a vertical tunnel or penstock. Many existing pumped hydroelectric consist of separate pump and turbine systems, but current configurations are based on reversible turbines. A separate pump and turbine system allows for a shorter transition time between pumping and generation modes, but its cost is high.

The model of a pumped hydroelectric requires considering the dynamics of the reservoir, just as in the case of hydraulic chains.

$$v_{t+1}^{up} = v_t^{up} - q_t \tag{9.12}$$

$$v_{t+1}^{dw} = v_t^{dw} + q_t \tag{9.13}$$

The model must include inflows and spillage in case they exist. The model must consider the net efficiency η for a charge/discharge cycle as presented in Equation (9.14).

$$p_t^{gen} - \eta p_t^{pmp} == \mu q_t \tag{9.14}$$

where p^{gen} is the power generated by the pumped hydroelectric and p^{pmp} is the power required from the system to pump water to the upper reservoir. Pumping requires more energy than is obtained by generating and hence $\eta \leq 1$. Equation (9.14) is valid only if p^{gen} is not positive simultaneously. That is to say the system is generating or pumping but not the two at the same time. This condition can be added to the model as the following set of constraints:

$$0 \leq p_t^{gen} \leq p_t^{max} x_t$$
$$0 \leq p_t^{pmp} \leq p_t^{max}(1 - x_t) \tag{9.15}$$

where x_t is a boolean variable. When $x_t = 0$ the generated power is zero and p^{pmp} takes values from zero to its maximum, the opposite occurs when $x_t = 1$.

The conventional use of pumped hydroelectric balances the load allowing nuclear plants to maintain constant power and/or to compensate for the high variability of wind and solar systems, as presented in the following example.

Example 9.4. Let us consider a generation system consisting of a large solar farm (100MW), a small thermal unit (10MW), and a pumped hydroelectric (30MW/120MWh). The system can buy and sell energy to the main grid; the objective is to maximize total income. Therefore, the pumped hydroelectric can buy energy from the grid at periods of low price to sell this energy at periods of a high price. The system is also able to store the energy generated by the solar plant. The price of the energy c_t is variable according to the hour, and the operation costs of the thermal unit are assumed linear. Therefore, the objective function is as follows:

$$\max f = \sum_{t \in T} c_t p_t - \alpha p_t^{thm} \tag{9.16}$$

where α is the incremental cost of the thermal unit, and p_t^{thm} is generated power at the time t. Moreover, p_t is the total power trade with the main grid, that is to say:

$$p_t = p_t^{thm} + p_t^{sol} + p_t^{gen} - p_t^{pmp} \tag{9.17}$$

where p^{gen} is the power injected by the hydroelectric in generation mode, p^{pmp} is the power taken from the grid in pumping mode, and p^{sol} is the power generated by the solar farm; notice that p_t may be negative, meaning the system is taking energy from the main grid to pump water.

The optimization model implemented in Python is presented below:

```
import cvxpy as cvx
pS = [0,0,0,0,0,0,0,26,50,71,87,97,100,97,87,71,50,26,0,0,0,0,0,0]
c  = [0.4, 0.4, 0.4, 0.4, 0.4, 0.5, 0.6, 0.6, 0.6, 0.5, 0.5, 0.4,
      0.4, 0.4, 0.5, 0.5, 0.6, 0.9, 1.1, 1.1, 1.0, 0.8, 0.7, 0.5]
```

```
vup = cvx.Variable(25,nonneg=True)
vdw = cvx.Variable(25,nonneg=True)
pgen = cvx.Variable(24,nonneg=True)
ppmp = cvx.Variable(24, nonneg=True)
pthm = cvx.Variable(24, nonneg=True)
q = cvx.Variable(24)
p = cvx.Variable(24)
x = cvx.Variable(24, boolean=True)
f = 0
res = [vup[0] == 0, vdw[0] == 120] # initial conditions
for t in range(24):
  f = f + c[t]*p[t]-0.95*pthm[t]
  res += [vup[t] <= 120, vdw[t] <= 120]
  res += [pgen[t] - 0.8*ppmp[t] == 1*q[t]]
  res += [pgen[t] - ppmp[t] + pS[t] + pthm[t] == p[t]]
  res += [pgen[t] <= 30*x[t], ppmp[t] <= 30*(1-x[t])]
  res += [vup[t+1] == vup[t] - q[t] ]
  res += [vdw[t+1] == vdw[t] + q[t] ]
  res += [pthm[t] <= 10]
PHS = cvx.Problem(cvx.Maximize(f),res)
PHS.solve()
print('eff:',print(np.sum(pgen.value-psto.value)))
```

Results of this problem are shown in Figure 9.6. The lower reservoir starts full and the upper reservoir empty. Prices are low in the first four hours, and hence, the hydroelectric unit starts pumping water; from 4 am to 9 am prices increase, making it viable to generate this energy stored; At medium day, solar generation is maximum, but prices are minimum. Therefore, it is convenient to store this energy pumping water to the upper reservoir; this energy is released to the grid from 16h to 20h where the prices are maximum. The thermal unit is turned on in this last period. The storage system ends with the same starting conditions (lower reservoir full and upper reservoir empty). Total efficiency of the storage process can be calculated by adding $p^{gen} - p^{pmp}$, in this case, the result is 40MW.

9.6 Further readings

The hydrothermal schedule has been usually solved by classic techniques such as linear programming, Lagrangian relaxation [61], and dynamic programming [56]. There is a vast literature about the use of metaheuristic techniques, such as simulated annealing [62] and genetic algorithms [63]. However, modern approaches are based on convex optimization, including semidefinite programming [54] and second-order cone approximations [55]. Other renewable generation can also be introduced in the model using stochastic optimization as presented in [64].

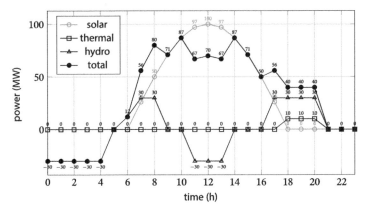

Figure 9.6 Results for the Example 9.4.

All models presented in this chapter simplify real operation problems, which can consider coupling with other models such as the unit commitment [65]. There is a vast literature in the field, especially in the power system society of IEEE; however, the problem has been studied by other communities, for example, the operation research community [66]. The problem may be extended to the operation planning that includes periods of one or several years; in that case, the problem is also stochastic. A tutorial on stochastic programming to solve this problem can also be found in [67].

9.7 Exercises

1. Solve the problem given in Example 9.1 considering the grid depicted in Figure 9.7. Use the transportation model with $p^{\max} = 150\text{MW}$ in all transmission lines.
2. Solve the hydrothermal scheduling problem given in Example 9.1 but now, consider a non-linear model of the hydroelectric power given by Equation (9.18),

$$p_H = h(v,q) = -0.0042v^2 + 0.03vq - 0.42q^2 + 0.9v + 10q - 50 \quad (9.18)$$

where v is the volume of the reservoir and q is the water discharge. Plot the surface and transform the equation into a second-order inequality constrain $h(v,q) \geq p_H$. Solve the corresponding hydrothermal dispatch and compare results with the linear model.
3. Solve the problem presented in Example 9.2 but without considering losses. Analyze and compare results.

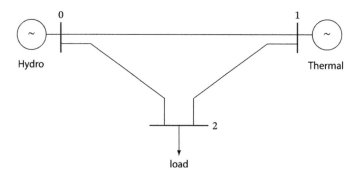

Figure 9.7 Three node system for hydrothermal scheduling.

4. Quadratic equality constraints related to power loss and water discharge were relaxed to convex inequality constraints in Example 9.2. Evaluate the accuracy of this approximation.
5. Execute the script presented in Example 9.3. Plot volume, water discharge, spillage, and generated power in each unit vs time.
6. Solve the hydrothermal scheduling problem given in Example 9.3 but assume that each hydroelectric unit is independent, i.e., without the hydraulic chain; compare results.
7. Solve the hydrothermal dispatch problem given in Example 9.3 considering time delays in the hydraulic chain. Consider $\tau_1 = \tau_2 = 1$ and $\tau_3 = 2$.
8. Solve the hydrothermal scheduling problem with pumped hydroelectric storage presented in Example 9.4 without allowing charge from the grid.
9. Solve the problem presented in Example 9.4 without considering the thermal unit.
10. Introduce a pump hydroelectric into Example 9.1. Use the parameters of Example 9.4.

10

Optimal power flow

Learning outcomes

By the end of this chapter, the student will be able to:

- Formulate the optimal power flow problem.
- Solve linear, SOC, and SDP approximations for the OPF.
- Identify the advantages and disadvantages of each approximation.

10.1 OPF in power distribution grids

Modern power distribution grids include renewable energy sources and energy storage devices that inject active and reactive power to the grid – each configuration of generation and demand results in a different operation point. However, not all operation points are equal; in practice, we seek operation points with minimum losses. This task is the main objective of the OPF.

A power distribution grid is represented as an oriented graph $\mathcal{G} = \{\mathcal{N}, \mathcal{E}\}$ where $\mathcal{N} = \{0, 1, 2, \ldots, n - 1\}$ is the set of nodes and $\mathcal{E} \subseteq \mathcal{N} \times \mathcal{N}$ is the set of edges. As convention, the slack node is 0 and its voltage is $v_0 = 1\angle 0$. The nodal admittance matrix is represented by $Y_{\text{bus}} = [y_{km}] \in \mathbb{C}^{n \times n}$ allowing to calculate nodal current from nodal voltages as given in (10.1).

$$i_k = \sum_{m \in \mathcal{N}} y_{km} v_m, \quad \forall k \in \mathcal{N} \tag{10.1}$$

This is an affine equation, thereby easily included in any optimization model. However, loads and generators are usually represented in terms of active and reactive power. Therefore, nodal equations become non-linear as given in (10.2).

Mathematical Programming for Power Systems Operation: From Theory to Applications in Python. First Edition. Alejandro Garcés.
© 2022 by The Institute of Electrical and Electronics Engineers, Inc. Published 2022 by John Wiley & Sons, Inc.

$$\left(\frac{s_k - d_k}{v_k}\right)^* = \sum_{m \in \mathcal{N}} y_{km} v_m, \ \forall k \in \mathcal{N} \tag{10.2}$$

where $(\cdot)^*$ represents the convex conjugate, s_k is the generated nodal power, and d_k is the corresponding load. For the sake of a compact representation of the model, we will assume that subscripts m and k belong to $\mathcal{N}$ in all cases. The model is presented in complex variable, for example, Equation (10.2) is represented in the complex domain; this is only a representation since the equation requires to be separated into real and imaginary parts. However, a complex representation is more direct when implemented in Python[1].

Although the problem may consider different objectives and may combine problems such as the economic/environmental dispatch, the typical application consists in minimizing power losses given by (10.3):

$$p_L = \text{real}\left(\sum_k \sum_m y_{km} v_k v_m^*\right) \tag{10.3}$$

This equation can be represented in a real domain by splitting $y_{km} = g_{km} + jb_{km}$ and $v = v^{\text{real}} + jv^{\text{imag}}$, resulting in the following equivalent expression:

$$p_L = \sum_k \sum_m g_{km}(v_k^{\text{real}} v_m^{\text{real}} + v_k^{\text{imag}} v_m^{\text{imag}}) \tag{10.4}$$

Since $G = [g_{km}] \in \mathbb{R}^{n \times n}$ is positive semidefinite[2], then p_L is a convex function. Thus, the basic model for the OPF is the following:

$$\min \ \text{real}\left(\sum_k \sum_m y_{km} v_k v_m^*\right)$$

$$v_0 = 1 + j0$$

$$1 - \delta \leq \|v_k\| \leq 1 + \delta, \ \forall k \in \mathcal{N}$$

$$p_k^{\max} \geq s_k^{\text{real}} \geq p_k^{\min}, \ \forall k \in \mathcal{N} \tag{10.5}$$

$$s_k^{\max} \geq \|s_k\|, \ \forall k \in \mathcal{N}$$

$$i_{km}^{\max} \geq \|y_{km}(v_k - v_m)\|, \ \forall (km) \in \mathcal{E}$$

$$\left(\frac{s_k - d_k}{v_k}\right)^* = \sum_m y_{km} v_m \ \forall k \in \mathcal{N}$$

---

1 See Section 4.6 in Chapter 4 for more details about optimization on the complex field.
2 This can be easily demonstrated taking into account that G can be calculated as $G = AG_p A^\top$, where A is the incidence matrix and G_p is a diagonal matrix with the resisting effect of each branch.

As we have seen, the objective function is convex; the first constraint is affine and represents the voltage in the slack node; the right-hand side of the second constraint is a second-order cone that represents the maximum deviation of the nodal voltage, whereas the left-hand side is a non-convex constraint that represents the minimum deviation of the nodal voltage. The value of the deviation δ is usually between 0.05pu to 0.10pu, according to the grid code in each country. The third and fourth constraints are the maximum capacity of each renewable source; the fifth constraint represents the thermal limit of each line, and the final constraint is the set of power flow equations. The latter is the primary source of complexity of this model; since it is not non-convex, therein lies the necessity of convex approximations to the OPF.

10.1.1 A brief review of power flow analysis

Before presenting convex approximations to the OPF, let us review some basic concepts from power flow analysis. First, it is important to differentiate the power flow analysis from the OPF. The former is the solution of a set of equations whereas the later is an optimization problem. The power flow problem allows to calculate nodal voltages from information of nodal powers. Since we know the voltage in the slack node ($v_0 = 1 + j0$), then we can divide the set of nodes as $\mathcal{N} = \{0, N\}$, where N are the nodes were the voltage is unknown. Therefore, the nodal admittance matrix can be represented as follows[3]:

$$Y_{\text{bus}} = \begin{pmatrix} Y_{00} & Y_{0N} \\ Y_{N0} & Y_{NN} \end{pmatrix} \tag{10.6}$$

With a slight abuse of notation, we can represent (10.2) in matrix form as given below:

$$\left(\frac{S_N - D_N}{V_N} \right)^* = v_0 Y_{N0} + Y_{NN} V_N \tag{10.7}$$

where $V_N = (v_1, v_2, \dots)^\top$ and $S_N = (s_1, s_2, \dots)^\top, D_N = (d_1, d_2, \dots)^\top$ are column vectors[4]. This is a set of non-linear algebraic equations that require a numerical method to find the value of V_N, a problem that can be solved by different methods such as Newton's and GaussŬ-Seidel. Here, we present a method based on a fixed point iteration, which is similar to the Gauss–Seidel method with a simple implementation in Python. Let us define the impedance matrix $Z_{NN} = Y_{NN}^{-1}$; this inverse exists as long as the graph that represents the grid is connected, which is the usual case; then Equation (10.7) can be represented as

---

3 See Appendix A for more details about the construction of the admittance matrix.
4 S/V indicates a division term to term.

the following fixed point:

$$V_N = T(V_N) \tag{10.8}$$

where T is a non-linear map from $\mathbb{C}^n$ to $\mathbb{C}^n$ given by Equation (10.9).

$$V_N = Z_{NN}\left(\left(\frac{S_N - D_N}{V_N}\right)^* - v_0 Y_{N0}\right) \tag{10.9}$$

The algorithm departs from an initial point $V_N = \mathbf{1}_N$ where $\mathbf{1}_N$ is a column vector with entries equal to one. Then, we evaluate $V_N \leftarrow T(V_N)$, and this iteration is repeated until achieving a fixed point, i.e., a point where $T(V_N) = V_N$. This is a solution to the set of algebraic equations. Although this is not the most efficient method to calculate a load flow, it is enough for our purposes, and, as we will see later, it is straightforward to implement. It is important to notice that a system may lack a solution or have several fixed points, some of them without practical meaning. However, under certain conditions, we can guarantee convergence and uniqueness of the solution with this approach, as was demonstrated in [68] for DC grids. Formal conditions for convergence and uniqueness of the solution are beyond the objectives of this book. Our approach is practical-oriented, and hence, convergence is checked by executing the algorithm.

Figure 10.1 shows a simple power distribution that will be used in later examples. These examples serve two purposes: first, to familiarize the reader with the implementation of graphs in Python, and second, to show the implementation of the power flow algorithms. This system will be used later in the OPF models, so it is recommended to implement and understand the following examples before continuing with subsequent sections.

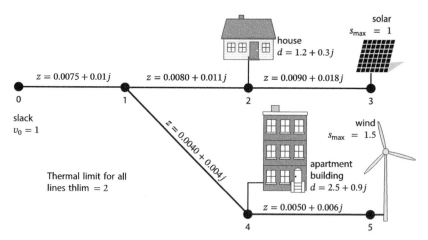

Figure 10.1 Example of a power distribution grid with distributed resources.

Example 10.1. All the information related to the power distribution grid depicted in Figure 10.1 can be stored in a single variable using the module networkx as presented below:

```
import numpy as np
import networkx as nx
G = nx.DiGraph()
G.add_node(0,name='slack',smax=10,d=0)
G.add_node(1,name='step',smax=0,d=0)
G.add_node(2,name='house',smax=0,d=1.2+0.3j)
G.add_node(3,name='solar',smax=1,d=0)
G.add_node(4,name='building',smax=0,d=2.5+0.9j)
G.add_node(5,name='wind',smax=1.5,d=0)
G.add_edge(0,1,y=1/(0.0075+0.010j),thlim=2)
G.add_edge(1,2,y=1/(0.0080+0.011j),thlim=2)
G.add_edge(2,3,y=1/(0.0090+0.018j),thlim=2)
G.add_edge(1,4,y=1/(0.0040+0.004j),thlim=2)
G.add_edge(4,5,y=1/(0.0050+0.006j),thlim=2)
nx.draw(G,with_labels=True,pos=nx.spectral_layout(G))
```

All the examples below depart from this definition of the graph, stored in a variable *G*. More details of this module are presented in Appendix A.

Example 10.2. We require to build the Y_{bus} as the block matrices given in (10.6). The nodal admittance matrix is calculated as given in (10.10):

$$Y_{bus} = AY_pA^T \tag{10.10}$$

where A is the incidence matrix of the oriented graph and Y_p is a diagonal matrix of the branch admittance. The incidence matrix can be easily obtained using the module networkx named as nx in Example 10.1, see the code below:

```
A = nx.incidence_matrix(G,oriented=True)
Yp = np.diag([G.edges[k]['y'] for k in G.edges])
Ybus = A@Yp@A.T
print(Ybus)
print(np.linalg.eigvals(Ybus.real))
```

In the last line, we checked if the real part of this matrix is positive semidefinite by calculating its eigenvalues.

Block matrices given in (10.6) are calculated from the Y_{bus} as follows:

```
n = G.number_of_nodes()
YN0 = Ybus[1:n,0]
YNN = Ybus[1:n,1:n]
ZNN = np.linalg.inv(YNN)
d = np.array([G.nodes[k]['d'] for k in G.nodes])
print(YN0)
print(YNN)
```

Example 10.3. The power flow equations seen as fixed point (10.8) allow a simple algorithm for calculating the operation point of the system. Let us define a function for the load flow calculation using this fixed point map with ten iterations:

```python
def LoadFlow(sN,dN):
    v0 = 1+0j
    vN = np.ones(n-1)*v0
    for t in range(10):
        vN = ZNN@(np.conj((sN-dN)/vN)-v0*YN0)
    vT = np.hstack([v0,vN]);
    sT = vT*np.conj(Ybus@vT)
    err = np.linalg.norm(sT[1:n]-(sN-dN))
    print('Load Flow, after 10 iterations the error is',err)
    return vT
```

The algorithm depart from $V_N = \mathbf{1}_N$ and evaluates the fixed point map (10.9). After that, the new voltages are stored in a variable V_T, including the slack node. Total loss is displayed at the end of the process. The algorithm can be improved using a while-loop instead of a for-loop (in this example, we preferred a compact code over an efficient algorithm).

We can evaluate the function using results from Example 10.2, considering loads exclusively (i.e., the solar panel and the wind turbine have generation equal to zero):

```python
VT = LoadFlow(np.zeros(n-1),d[1:n])
ST = VT*np.conj(Ybus@VT)
pL = sum(ST)
print('Loss',pL)
for (k,m) in G.edges:
    Sf = Ybus[k,m]*(VT[k]-VT[m])
    print('flow',(k,m),np.abs(Sf))
```

Results can be stored in a DataFrame as follows:

```python
import pandas as pd
results = pd.DataFrame()
results['name'] = [G.nodes[k]['name'] for k in G.nodes]
results['vpu'] = np.abs(VT)
results['ang'] = np.angle(VT)*180/np.pi
results['pnode'] = np.round(ST.real,4)
results['qnode'] = np.round(ST.imag,4)
results.head(n)
```

The reader can verify that nodal voltages are $v = (1, 0.956, 0.943, 0.943, 0.943, 0.943)^\mathsf{T}$ and power loss is $p_L = 0.173$.

Example 10.4. Node 1 in the system depicted in Figure 10.1 does not have generation or load. Therefore, it can be eliminated using a Kron reduction. Let us split the set of nodes $\mathcal{N} = \{s, r\}$ where s is the set of nodes with nodal current equal to zero, and r are the remaining nodes. Then, we have the following:

$$0 = Y_{ss}V_s + Y_{sr}V_r \tag{10.11}$$

$$I_r = Y_{rs}V_s + Y_{rr}V_r \tag{10.12}$$

where $Y_{ss}, Y_{sr}, Y_{rs}, Y_{rr}$ are block matrices from Y_{bus}. Therefore, we can define a reduced admittance matrix Y_{kron} as follows:

$$Y_{\text{kron}} = Y_{rr} - Y_{rs}Y_{ss}^{-1}Y_{sr} \tag{10.13}$$

This equation can be coded in Python as presented below for a single node $s = [1]$:

```
s = [1]
r = list(set(range(n)).difference(s))
nn = len(r)
Ykron = np.zeros((5,5))*0j
for k in range(nn):
  for m in range(nn):
    Ykron[k,m] = Ybus[r[k],r[m]]-1/Ybus[s,s]*Ybus[r[k],s]*
    Ybus[s,r[m]]
```

Kron reduction is used extensively in many power systems applications, and therefore, it is useful to have this code for future examples.

10.2 Complex linearization

As we have seen, the fundamental OPF problem given by (10.5) is non-convex due to the power flow equations, and hence, a convex approximation is required. In this section, we present a simple linearization based on Wirtinger calculus. There are many other linearizations in the literature (most of them equivalent), but the representation presented here has advantages in terms of accuracy and simple implementation.

Let us represent (10.2) as the following equivalent algebraic system:

$$s_k^* - d_k^* = \sum_m y_{km} w_{km} \tag{10.14}$$

$$w_{km} = v_k^* v_m \tag{10.15}$$

where w_{km} is a new complex variable[5]. Notice that Equation (10.14) is affine and the non-convexity appears in Equation (10.15). This equations can be linearized in the complex plain around a given point u_k, u_m, using Wirtinger's calculus (see Appendix B for more details):

$$w_{km} - u_k^* u_m = u_k^*(v_m - u_m) + u_m(v_k^* - u_k^*) \tag{10.16}$$

usually $u_k = u_m = 1\mathrm{pu}$ resulting in the following affine constraint:

$$w_{km} = v_k^* + v_m - 1 \tag{10.17}$$

This simple equation constitutes a convex linearization of the power flow equations.

On the other hand, voltage limitation introduces another set of non-convex constraints, namely:

$$1 - \delta \le \|v_k\| \le 1 + \delta \tag{10.18}$$

The right hand is a ball or radius $1 + \delta$ which is, of course, convex. However, the left-hand side is a non-convex constraint since it is the exterior of a open ball of radius $1 - \delta$. The set defined by (10.18) (both left- and right-hand sides) is named an annulus and is a non-convex set. In practice, this set can be replaced by the following set:

$$1 - \delta \le v_k^{\mathrm{real}} \le 1 + \delta \tag{10.19}$$

$$1 - \delta \le v_k^{\mathrm{imag}} \le 1 + \delta \tag{10.20}$$

or equivalently as:

$$\|v_k - 1\|_1 \le \delta \tag{10.21}$$

where $\|\cdot\|_1$ represents the 1-norm. Although this approximation may seem arbitrary, the following example shows in logic behind it.

Example 10.5. Constraint (10.21) is suitable approximation for values of $\delta = 0.1$ and below. Figure 10.2 shows a subsection of the annulus $0.9 \le \|v_k\| \le 1.1$ for values around $1 + 0j$. The box constraint (10.21) is represented as a shadow square which is a close approximation of the set for angles between $\theta = \pm 5^o$ and $\theta = \pm 7^o$. Voltage angles are usually small in power distribution networks [69], so this is a fair approximation. The model may be complemented by constraints on the angle, which are convex. An exact value for the maximum angle would require stability criteria beyond the objectives of this book. A more conservative constraint is obtained by replacing the 1-norm with a 2-norm in (10.21).

---

5 This new variable increases the dimension of the set of feasible solutions; sometimes the nature of the problem is only revealed when we change our perspective to a higher dimension.

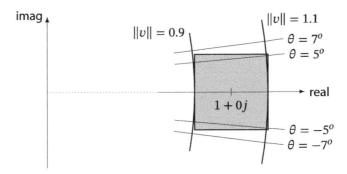

Figure 10.2 Approximation of the voltage restriction as a box constraint.

Combining the aforementioned approximations, the OPF is transformed into the following convex optimization problem:

$$\min \ \text{real}\left(\sum_k \sum_m y_{km} v_k v_m^*\right)$$

$$v_0 = 1 + j0$$

$$\delta \geq \|v_k - 1\|_1 , \ \forall k \in \mathcal{N}$$

$$p_k^{\max} \geq \text{real}(s_k) \geq p_k^{\min}, \ \forall k \in \mathcal{N} \tag{10.22}$$

$$s_k^{\max} \geq \|s_k\| , \ \forall k \in \mathcal{N}$$

$$i_{km}^{\max} \geq \|y_{km}(v_k - v_m)\| , \ \forall (km) \in \mathcal{E}$$

$$s_k^* - d_k^* = \sum_m y_{km} w_{km}, \ \forall k \in \mathcal{N}$$

$$w_{km} = v_k^* + v_m - 1, \ \forall (km) \in \mathcal{N} \times \mathcal{N}$$

notice that w_{km} increases the number of variables of the model; however, the new equations are affine and hence, it is not a problem in practice. It is also possible to replace (10.17) into (10.14) to obtain a model with the same number of variables as the original problem. Here, we are prioritizing a simple representation over the efficiency of the algorithm.

Notice the model is still non-linear since the objective function is quadratic. In addition, there are second-order constraints related to the nodal voltage and the capacity of each renewable power resource. However, these non-linear equations generate a convex model that can be efficiently solved using CvxPy, as shown in the following example.

Example 10.6. Let us solve the OPF problem for the system given in Figure 10.1 using a convex linearization of the power flow equations. We assume we have stored the graph as given in Example 10.1 and calculated the Y_{bus} as shown in Example 10.2. Both solar panel and wind turbine are available to generate its nominal power. The code is presented below:

```
import cvxpy as cvx
smax = np.array([G.nodes[k]['smax'] for k in G.nodes])
d = np.array([G.nodes[k]['d'] for k in G.nodes])
v = cvx.Variable(n,complex=True)
s = cvx.Variable(n,complex=True)
W = cvx.Variable((n,n),complex=True)
obj = cvx.Minimize(cvx.quad_form(cvx.real(v),Ybus.real)+
                   cvx.quad_form(cvx.imag(v),Ybus.real))
res = [v[0] == 1.0]
M = Ybus@W
for k in G.nodes:
  res += [cvx.conj(s[k]-d[k]) == M[k,k]]
  res += [cvx.abs(v[k]-1) <= 0.05]
  res += [cvx.abs(s[k]) <= smax[k]]
  for m in G.nodes:
    res += [W[m,k] == cvx.conj(v[k])+v[m]-1]
for (k,m) in G.edges:
  res += [cvx.abs(Ybus[k,m]*(v[k]-v[m])) <= G.edges[(k,m)]
  ['thlim']]
OPF = cvx.Problem(obj,res)
OPF.solve()
print('pL',obj.value,OPF.status)
```

Most of the lines in this code are self explanatory; however, there are some aspects that require careful explanation. First, notice that $W = [w_{km}]$ is a matrix of the same size of Y_{bus}, therefore, we can define a new matrix given by (10.23):

$$M = Y_{bus}W \tag{10.23}$$

This matrix allows to represent (10.14) as follows:

$$s_k^* - d_k^* = m_{kk} \tag{10.24}$$

Second, set representations such as $\forall k \in \mathcal{N}$ help to define the for-loop in the code. So, $\forall(km) \in \mathcal{N} \times \mathcal{N}$ indicates a nested-loop whereas $\forall(km) \in \mathcal{E}$ indicates a for-loop in the set of the edges. In this case, $\mathcal{N}$ is equivalent to G.nodes (we can also use range(n)), and $\mathcal{E}$ is equivalent to G.edges.

The reader can prove that the result of this problem is $s_3 = 0.96 + 0.28j$ and $s_5 = 1.42 + 0.49j$, with $p_L = 0.0406$. However, this is an approximation of the power loss which requires to be calculated via a power flow analysis:

```
VT = LoadFlow(s.value[1:n],d[1:n])
ST = VT*np.conj(Ybus@VT)
pL = sum(ST)
print('Loss',pL)
```

After executing this code, power loss is $p_L = 0.0406$ (a great reduction in comparison to Example 10.3). Notice that although the solar panel has a capacity of $s_k^{\max} = 1$, not all generation is an active power. The algorithm chooses to reduce its active power in order to generate some reactive power and minimize power loss. In case the primary resource (i.e., wind/solar) is limited, then we require to include constraints of the form $\text{real}(s_k) \leq p_k^{\max}$. This constraint is, of course, affine and does not represent a complication of the model. The reader is invited to compare nodal voltages in the system, with and without distributed generation.

10.2.1 Sequential linearization

We can improve the results of the linearization by linearizing again in the new operating point. The algorithm is quite simple; we start with a vector $U = 1_N$ and linearize the power flow equations around this point. Then we solve Model (10.22) obtaining a vector S with the power generated by each unit. Then, we calculate a power flow, using, for instance, the fixed-point algorithm given in Example 10.3. This algorithm returns a new set of voltages U, which are used to linearize the model again using (10.16). The optimization model is again solved using this new linearization, and the steps are repeated until achieving convergence.

This method does not guarantee global optimum, but it is efficient in practice, as shown in the following example:

Example 10.7. In the following code, we make three iterations of sequential linearizations in order to obtain a better approximation of the optimal solution. First, we define a function named `LinearOPF` which solves the optimization model for a linearization around a point V_T:

```
def LinearOPF(u):
    v = cvx.Variable(n,complex=True)
    s = cvx.Variable(n,complex=True)
    W = cvx.Variable((n,n),complex=True)
    obj = cvx.Minimize(cvx.quad_form(cvx.real(v),Ybus.real)+
                       cvx.quad_form(cvx.imag(v),Ybus.real))
    M = Ybus@W
    res = [v[0] == 1.0]
```

```
for k in range(n):
    res += [cvx.conj(s[k]-d[k]) == M[k,k]]
    res += [cvx.abs(s[k]) <= smax[k]]
    for m in range(n):
        res += [W[m,k] == cvx.conj(v[k])*u[m]+
                          np.conj(u[k])*v[m]
                          -np.conj(u[k])*u[m]]
OPF = cvx.Problem(obj,res)
OPF.solve()
print('pL',obj.value,OPF.status)
return s.value
```

The main difference of this model with respect to the model in Example 10.6 is the point in which w_{km} is linearized; in this case, we linearize around U. Next, we evaluate this function as well as the power flow already defined in Example 10.3, namely:

```
VT = np.ones(n)*(1.0+0.0j)
for t in range(3):
    ST = LinearOPF(VT)
    VT = LoadFlow(ST[1:n],d[1:n])
    print('Loss',sum(VT*np.conj(Ybus@VT)))
```

Power loss is $p_L = 0.04258$ for both the load flow and the linear OPF.

10.2.2 Exponential models of the load

Loads in power distribution grids are usually represented as exponential models as presented below:

$$d_k = d_k^{\text{real}} \|v_k\|^{\alpha} + j d_k^{\text{imag}} \|v_k\|^{\beta} \tag{10.25}$$

where α, β are real numbers that represent the variation of the active and reactive power, with respect to the voltage. Typical values of α, β are $\alpha = \beta = 0$ for industrial loads, $\alpha = \beta = 1$ for commercial loads and $\alpha = \beta = 2$ for residential loads. Nevertheless, fractional values are allowed.

Equation (10.25) leads to a non-convex constraint, however, it can be easily linearized using Wirtinger's calculus. We present only the linearization of $\|v_k\|^{\alpha}$ since the linearization of $\|v_k\|^{\alpha}$ follows the same procedure. First, consider the following complex function:

$$\|v\|^{\alpha} = (vv^*)^{\alpha/2} \tag{10.26}$$

then, we linearize this equation by derivating with respect to v and v^* and evaluating in a reference value v_0:

$$\|v\|^\alpha \approx (v_0 v_0^*)^{\frac{\alpha}{2}} + \frac{\alpha}{2}(v_0 v_0^*)^{\frac{\alpha}{2}-1} v_0^* \Delta v + \frac{\alpha}{2}(v_0 v_0^*)^{\frac{\alpha}{2}-1} v_0 \Delta v^* \tag{10.27}$$

$$= (v_0 v_0^*)^{\frac{\alpha}{2}} + \frac{\alpha}{2}(v_0 v_0^*)^{\frac{\alpha}{2}-1} v_0^*(v-v_0) + \frac{\alpha}{2}(v_0 v_0^*)^{\frac{\alpha}{2}-1} v_0(v^*-v_0^*) \tag{10.28}$$

$$= a + bv + b^* v^* \tag{10.29}$$

with

$$a = (1 - \alpha)\|v_0\|^\alpha \tag{10.30}$$

$$b = \frac{\alpha}{2}(v_0 v_0^*)^{\frac{\alpha}{2}-1} v_0^* \tag{10.31}$$

For the case of $v_0 = 1 + 0j$, the linearization is simplified as (10.32):

$$\|v\|^\alpha \approx 1 - \alpha + \frac{\alpha}{2}(v + v^*) \tag{10.32}$$

Example 10.8. We are going to evaluate the accuracy of Equation (10.32) for voltages close to 1pu. Let us define the following functions:

$$f(v) = \|v\|^\alpha \tag{10.33}$$

$$g(v) = 1 - \alpha + \frac{\alpha}{2}\left(v_k + v_k^*\right) \tag{10.34}$$

$$\epsilon(v) = \|f(v) - g(v)\| \tag{10.35}$$

where $f : \mathbb{C} \to \mathbb{R}$ is the exact exponential function, $g : \mathbb{C} \to \mathbb{R}$ is its linearization, and $\epsilon : \mathbb{C} \to \mathbb{R}$ is the total error. We evaluate this error in $n = 10^4$ random points generated in set Ω presented below:

$$\Omega = \left\{v \in \mathbb{C} : 0.9 \le v^{\text{real}} \le 1.1, -0.1 \le v^{\text{imag}} \le 0.1\right\} \tag{10.36}$$

A distribution function is obtained which gives the expected error with a defined probability. The corresponding script in Python is presented below:

```
import numpy as np
import matplotlib.pyplot as plt
n = 10000
vreal = [0.9+0.2*np.random.rand() for k in range(n)]
vimag = [0.1-0.2*np.random.rand() for k in range(n)]
v = np.array(vreal) + 1j*np.array(vimag)
alpha = 2
f = np.abs(v)**alpha
g = 1-alpha + alpha/2*(v+v.conj())
error = np.abs(f - g)
```

```
error.sort()
probability = np.linspace(0,1,n)
plt.plot(error*100,probability)
plt.grid()
plt.show()
```

Figure 10.3 shows the results for $\alpha = 2$. This plot represents the cumulative distribution function of ϵ. In this case, 80% of the randomly generated points produced an error less than 1%.

This demonstrates the high accuracy of the method. The student is invited to generate this plot for different n and different values of $\alpha > 0$.

10.3 Second-order cone approximation

A Second-order cone approximation is a convenient manner to include power flow equations into an optimization model, especially for power distribution applications. In this case, we convexify the equations maintaining the non-linear nature of the problem. We depart from (10.15) which is the primary non-convex constraint in the problem. Let us multiply by w_{km}^* obtaining the following equivalent equation:

$$w_{km}w_{km}^* = (v_k^*v_m)(v_kv_m^*) \tag{10.37}$$

which can be written as

$$\|w_{km}\|^2 = \|v_k\|^2 \|v_m\|^2 \tag{10.38}$$

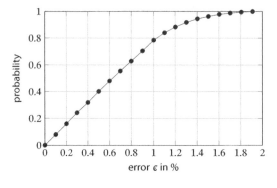

Figure 10.3 Cumulative distribution function of the linearization error for $\alpha = 2$

Let us define a new vector $H \in \mathbb{R}^n$ with entries $h_k = \|v_k\|^2$, then (10.38) is transformed into (10.39)

$$\|w_{km}\|^2 = h_k h_m \tag{10.39}$$

At this point, this constraint is still non-convex; therefore, we propose an approximation that consists in transforming the equality into inequality and solve the resulting hyperbolic set as previously presented in Example 5.3, Chapter 5, namely:

$$\left\| \begin{pmatrix} 2w_{km} \\ h_k - h_m \end{pmatrix} \right\| \leq h_k + h_m \tag{10.40}$$

The limit in each distribution line can be represented as function of the new variables h_k by multipling the branch current by v_k^* as follows:

$$i_{km} v_k^* = y_{km} v_k^* (v_k - v_m), \quad \forall (km) \in \mathcal{E} \tag{10.41}$$

which in turn is transformed into the following affine equation:

$$s_{km} = y_{km}(h_k - w_{km}) \tag{10.42}$$

For the sake of simplicity, we assume $i_{km}^{max} = s_{km}^{max}$ to complete the model; thus, the SOC approximation for the OPF is presented below:

$$\text{min real}\left(\sum_k s_k - d_k \right)$$

$$h_0 = 1$$

$$(1 + \delta)^2 \geq h_k \geq (1 - \delta)^2, \quad \forall k \in \mathcal{N}$$

$$p_k^{max} \geq \text{real}(s_k) \geq p_k^{min}, \quad \forall k \in \mathcal{N}$$

$$s_k^{max} \geq \|s_k\|, \quad \forall k \in \mathcal{N} \tag{10.43}$$

$$i_{km}^{max} \geq \|y_{km}(h_k - w_{km})\|, \quad \forall (km) \in \mathcal{E}$$

$$s_k^* - d_k^* = \sum_m y_{km} w_{km}, \quad \forall k \in \mathcal{N}$$

$$\left\| \begin{pmatrix} 2w_{km} \\ h_k - h_m \end{pmatrix} \right\| \leq h_k + h_m, \quad \forall (km) \in \mathcal{N} \times \mathcal{N}$$

In this model, we calculated power loss as the sum of nodal powers. This equation is entirely equivalent.

Example 10.9. Let us implement the SOC model given by (10.43) for the distribution system shown in Figure 10.1. The code in Python is given below:

```
smax = np.array([G.nodes[k]['smax'] for k in G.nodes])
d = np.array([G.nodes[k]['d'] for k in G.nodes])
h = cvx.Variable(n)
s = cvx.Variable(n,complex=True)
W = cvx.Variable((n,n),complex=True)
M = Ybus@W
res = [h[0] == 1.0]
for k in range(n):
  res += [cvx.conj(s[k]-d[k]) == M[k,k]]
  res += [cvx.abs(s[k]) <= smax[k]]
  res += [h[k] >= 0.9025]
  res += [h[k] <= 1.1025]
  res += [W[k][k] == h[k]]
  for m in range(n):
    res += [cvx.SOC(h[k]+h[m], cvx.vstack([2*W[k,m],
              h[k]-h[m]]))]
  res += [W[m,k] == cvx.conj(W[k,m])]
for (k,m) in G.edges:
  ylin = np.abs(G.edges[(k,m)]['y'])
  slin = G.edges[(k,m)]['thlim']
  res += [cvx.abs(h[k]-W[k,m]) <= slin/ylin]
  res += [cvx.abs(h[m]-W[m,k]) <= slin/ylin]
obj = cvx.Minimize(cvx.sum(cvx.real(s-d)))
OPFSOC = cvx.Problem(obj,res)
OPFSOC.solve()
print('pL',obj.value,OPFSOC.status)
```

After executing this code, power loss is $p_L = 0.04259$ with $s_3 = 0.9534 + 0.3018j$ and $s_5 = 1.4115 + 0.5076j$. The student may observe this solution is close to the solution obtained by the power flow analysis. 🐍

Example 10.10. The magnitude of nodal voltages can be recovered from the results of Model (10.43) without executing a new power flow analysis as follows:

$$v_k = \sqrt{h_k} \tag{10.44}$$

The angle can be calculated evaluating the angle of each edge of the graph as given below:

$$\theta_k - \theta_m = \text{angle}(w_{km}), \; \forall(km) \in \mathcal{E} \tag{10.45}$$

The reader is invited to evaluate these equations and compare them to the power flow results with the powers resulting from the SOC approximation. 🐍

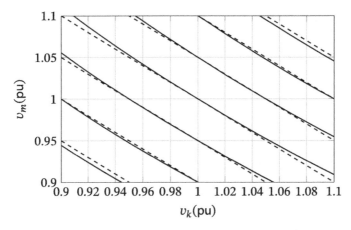

Figure 10.4 Comparison between the constraint $w_{km} = v_k + v_m - 1$ (solid line) and the constraint $w_{km}^2 \leq v_k^2 v_m^2$ (dashed line) for $(v_k, v_m, w_{km}) \in \mathbb{R}^3$.

Example 10.11. Both SOC approximation and linearization are based on Equation (10.15), which goes from $\mathbb{C}^2$ to $\mathbb{C}$. Although it is difficult to visualize a function in $\mathbb{C}^2$, it is possible to make a plot when it goes from $\mathbb{R}^2$ to $\mathbb{R}$ (that is the case of the OPF on DC grids). Figure 10.4 shows a comparison among the two approximations. The dashed line defines a linear approximation around 1pu, whereas the solid line defines a hyperbolic set that can be transformed into an SOC. Both approximations are quite similar, although the linear approximation is more imprecise as the voltages go far from 1pu. In practice, voltages are 1 ± 0.1 up so that the linear approximation is enough. It should be noted that the OPF constitutes the tertiary control in active distribution networks, and it requires to be evaluated in real-time. Therefore, we require to define a suitable tray-off between accuracy and speed. Linearization is perhaps the best approach in this case (see [70] for more details about the linear approximation).

Another way to visualize the difference between linear and SOC approximations is by generating random samples of v_k and v_m on the complex set Ω given by (10.36), and evaluate the error as in Example 10.8. The reader is invited to do this numerical experiment.

10.4 Semidefinite approximation

Semidefinite programming allows to generate a highly accurate approximation for the OPF problem. Unlike the approaches previously presented, in this case it is convenient to separate nodal voltages in real and imaginary parts as $v_k = f_k + je_k$ and define the following block matrix:

$$\begin{pmatrix} E & Z \\ Z^\mathsf{T} & F \end{pmatrix} \tag{10.46}$$

where the entries of each block matrix are defined as follows:

$$E_{km} = e_k e_m \tag{10.47}$$

$$Z_{km} = e_k f_m \tag{10.48}$$

$$F_{km} = f_k f_m \tag{10.49}$$

Notice that (10.46) is positive semidefinite and rank 1. Therefore it is possible to generate an SDP approximation of (10.5) by representing all the models as a function of this matrix, relaxing the rank constraint.

As we have seen, a key step is to find a suitable representation for w_{km} in (10.15), and in this case, it is easy to see the following:

$$w_{km} = E_{km} + F_{km} + j(Z_{km} - Z_{mk}) \tag{10.50}$$

Moreover, we can obtain a suitable representation of the objective function from (10.4),

$$p_L = \text{trace}(G_{\text{bus}}E + G_{\text{bus}}F) \tag{10.51}$$

where $G_{\text{bus}} = \text{real}(Y_{\text{bus}})$. With these simple change of variables, we obtain the following semidefinite problem:

$$\begin{aligned}
\min \; & \text{trace}(G_{\text{bus}}E + G_{\text{bus}}F) \\
& E_{00} = 1 \\
& F_{0k} = 0 \\
& s_k^* - d_k^* = \sum_m y_{km} w_{km} \\
& w_{km} = E_{km} + F_{km} + j(Z_{km} - Z_{mk})
\end{aligned} \tag{10.52}$$

$$\begin{pmatrix} E & Z \\ Z^{\mathsf{T}} & F \end{pmatrix} \succeq 0$$

$$\begin{pmatrix} s_k^{\max} & 0 & \text{real}(s_k) \\ 0 & s_k^{\max} & \text{imag}(s_k) \\ \text{real}(s_k) & \text{imag}(s_k) & s_k^{\max} \end{pmatrix} \succeq 0$$

For simplicity, we have omitted some constraints related to the thermal limit and voltage regulation. In addition, the last constraint which is equivalent to $\|s_k\| \leq s_k^{\max}$ is transformed into a semidefinite constraint in order to obtain a pure SDP problem.

Example 10.12. The code presented below represents a SDP approximation for the optimal power flow problem:

```
smax = np.array([G.nodes[k]['smax'] for k in G.nodes])
E = cvx.Variable((n,n),symmetric=True)
F = cvx.Variable((n,n),symmetric=True)
Z = cvx.Variable((n,n))
s = cvx.Variable(n,complex=True)
W  = cvx.Variable((n,n),complex=True)
obj = cvx.Minimize(cvx.trace(Ybus.real@E+Ybus.real@F))
M = Ybus@W
res = [E[0,0] == 1]
res += [cvx.bmat([[E,Z],[Z.T,F]]) >> 0]
res += [cvx.trace(Ybus.real@E+Ybus.real@F) == cvx.sum
(cvx.real(s-d))]
for k in range(n):
   res += [cvx.conj(s[k]-d[k]) == M[k,k]]
   res += [F[k,0] == 0]
   res += [F[0,k] == 0]
   res += [cvx.bmat([[smax[k],0,cvx.real(s[k])],
                     [0,smax[k],cvx.imag(s[k])],
                     [cvx.real(s[k]),cvx.imag(s[k]),
                     smax[k]]]) >> 0]
   for m in range(n):
      res += [W[k,m] == E[k,m] + F[k,m] + 1j*(Z[k,m]-Z[m,k])]
OPFSDP = cvx.Problem(obj,res)
OPFSDP.solve()
print('pL',obj.value,OPFSDP.status)
```

The results of this model evaluated in the distribution system depicted in Figure 10.1 are $p_L = 0.042587$, $s_{\text{solar}} 0.9534 + 0.3017j$, and $s_{\text{wind}} = 1.4115 + 0.5077j$.

10.5   Further readings

The OPF has been studied for a long time, with classic approaches based on non-linear programming as can be found in [71], [72], and [73]. The interior-point method seems to work well in practice, even for non-convex formulations, as demonstrated in [74]. However, these applications do not analyze characteristics such as convergence and global optima. Therefore, linearizations and cone approximations are required usually proposed.

Although there is a proliferation of linearizations for the power flow equations (most of them entirely equivalent), the approximations presented in this chapter are based on [75] and [70] that allow complex representations easily implementable in Python even for three-phase unbalanced power distribution systems.

There is also vast literature about second-order cone approximations, standing out the work of Low (see for example [76] and [77]). A complete review of the problem can be found in [78]. This review includes linearizations and cone approximations.

The OPF can be extended to DC grids in both high voltage and microgrids. The interested reader can refer to [79] for the case of microgrids.

Semidefinite programming has also been an active area of research for OPF problems [80]. A complete analysis of the geometry of the problem can be found in [81]. Most of the approximations presented in this chapter are suitable for radial distribution grids; an analysis for meshed grids can be found in [82]

10.6   Exercises

1. Make a comparative analysis among linearization, sequential linearization, SOC and SDP approximations. Identify the advantages and disadvantages of each approach.
2. Analyze convergence properties of the power flow algorithm presented in Section 10.1.1, by plotting a curve of error vs iterations for different values of load.
3. Compare numerical results for linearization, sequential linearization, SOC and SDP approximations in the system depicted in Figure 10.1 for different values of loads and power factor.

4. Calculate the OPF for the system depicted in Figure 10.1 if a new line is included between Node 3 and Node 5, with $z_{15} = 0.0060 + 0.01$.
5. Solve the OPF using the linearization and SOC approximation, for the power distribution system given in Table 10.1. Assume there is distributed generation at nodes 11, 20, 21, and 26 with nominal capacity $s^{max} = 0.08$pu.
6. T1, T2, and T3 in Table 10.1 represent the load curve for the 34-bus test system. T1 represents the load factor for operation between 0h:8h, T2 for 8h:16h, and T3 for 16h:24h. Solve the OPF problem for each of this operation points.
7. Conventional distributed generation are based on synchronous machines instead of power electric converters. This type of machine presents a capability curve as shown in Figure 10.5. Include this capability curve into the OPF model (notice this curve generates a convex constraint).
8. Include voltage constraints in the SDP approximation given in (10.52). Recover the nodal voltages from the semidefinite approximation presented in Example 10.12.
9. The optimal power flow problem can be extended to the operation of DC distribution systems. Consider the DC distribution system given in

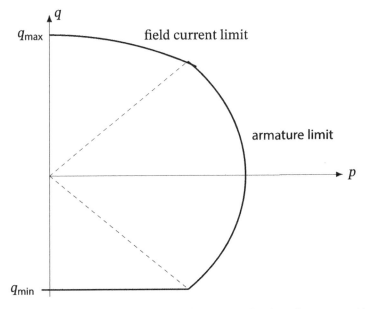

Figure 10.5 Example of a capability curve for a conventional synchronous machine.

Table 10.1 34-bus test system taken from [83].

From	To	r_{km}[pu]	x_{km}[pu]	p[pu]	q[pu]	T1	T2	T3
1	2	0.00967	0.00397	0.0230	0.01425	0.55	0.70	0.65
2	3	0.00886	0.00364	0.0000	0.00000	0.00	0.00	0.00
3	4	0.01359	0.00377	0.0230	0.01425	0.55	0.70	0.65
4	5	0.01236	0.00343	0.0230	0.01425	0.55	0.70	0.65
5	6	0.01236	0.00343	0.0000	0.00000	0.00	0.00	0.00
6	7	0.02598	0.00446	0.0000	0.00000	0.00	0.00	0.00
7	8	0.01732	0.00298	0.0230	0.01425	0.55	0.70	0.65
8	9	0.02598	0.00446	0.0230	0.01425	0.55	0.70	0.65
9	10	0.01732	0.00298	0.0000	0.00000	0.00	0.00	0.00
10	11	0.01083	0.00186	0.0230	0.01425	0.55	0.70	0.65
11	12	0.00866	0.00149	0.0137	0.00840	0.50	0.60	0.55
3	13	0.01299	0.00223	0.0072	0.00450	0.45	0.65	0.60
13	14	0.01732	0.00298	0.0072	0.00450	0.45	0.65	0.60
14	15	0.00866	0.00149	0.0072	0.00450	0.45	0.65	0.60
15	16	0.00433	0.00074	0.0014	0.00075	0.60	0.70	0.65
6	17	0.01483	0.00412	0.0230	0.01425	0.55	0.70	0.65
17	18	0.01359	0.00377	0.0230	0.01425	0.55	0.70	0.65
18	19	0.01718	0.00391	0.0230	0.01425	0.55	0.70	0.65
19	20	0.01562	0.00355	0.0230	0.01425	0.55	0.70	0.65
20	21	0.01562	0.00355	0.0230	0.01425	0.55	0.70	0.65
21	22	0.02165	0.00372	0.0230	0.01425	0.55	0.70	0.65
22	23	0.02165	0.00372	0.0230	0.01425	0.55	0.70	0.65
23	24	0.02598	0.00446	0.0230	0.01425	0.55	0.70	0.65
24	25	0.01732	0.00298	0.0230	0.01425	0.55	0.70	0.65
25	26	0.01083	0.00186	0.0230	0.01425	0.55	0.70	0.65
26	27	0.00866	0.00149	0.0137	0.00850	0.50	0.60	0.55
7	28	0.01299	0.00223	0.0075	0.00480	0.55	0.75	0.70
28	29	0.01299	0.00223	0.0075	0.00480	0.55	0.75	0.70
29	30	0.01299	0.00223	0.0075	0.00480	0.55	0.75	0.70
10	31	0.01299	0.00223	0.0057	0.00345	0.57	0.63	0.58
31	32	0.01732	0.00298	0.0057	0.00345	0.57	0.63	0.58
32	33	0.01299	0.00223	0.0057	0.00345	0.57	0.63	0.58
33	34	0.00866	0.00149	0.0057	0.00345	0.57	0.63	0.58

Table 10.2; solve the corresponding OPF problem using linearization and SOC approximation.

Table 10.2 Parameters of a 21-nodes DC power distribution grid

From	To	r[pu]	d[pu]	p^{max}[pu]
1	2	0.0053	0.70	0.0
1	3	0.0054	0.00	0.0
3	4	0.0054	0.36	0.0
4	5	0.0063	0.04	0.0
4	6	0.0051	0.36	0.0
3	7	0.0037	0.00	0.0
7	8	0.0079	0.32	0.0
7	9	0.0072	0.80	1.5
3	10	0.0053	0.00	0.0
10	11	0.0038	0.45	0.0
11	12	0.0079	0.68	1.5
11	13	0.0078	0.10	0.0
10	14	0.0083	0.00	0.0
14	15	0.0065	0.22	0.0
15	16	0.0064	0.23	0.0
16	17	0.0074	0.43	0.0
16	18	0.0081	0.34	1.5
14	19	0.0078	0.09	0.0
19	20	0.0084	0.21	0.0
19	21	0.0082	0.21	3.0

10. Equation (10.26) may be represented as follows:

$$\|v_k\|^{\alpha} = \left(\sqrt{x^2 + y^2}\right)^{\alpha} \tag{10.53}$$

with $x = v_k^{real}$ and $y = v_k^{imag}$. This is an equation from $\mathbb{R}^2$ to $\mathbb{R}$. Use a Taylor expansion to linearize this equation around $x = v_k^{real} = 1$ and $y = v_k^{imag} = 0$. Compare the result with (10.32).

11

Active distribution networks

Learning outcomes

By the end of this chapter, the student will be able to:

- Formulate mixed-integer convex models for the optimal placement of capacitors and distributed generation.
- Formulate a mixed-integer convex model for the optimal placement of distributed generation.
- Formulate a convex model for hosting capacity.

11.1 Modern distribution networks

Modern power distribution networks include several active elements, such as distributed generation and electric vehicles, that must be included in the operation via optimization models. Convex approximations for the optimal power flow equations previously presented in Chapter 10 are key to formulate these optimization models; therefore, it is convenient to review these formulations, especially the linear formulation before continuing with the sections below. We present three main problems, namely: optimal placement of capacitors, optimal placement and size of distributed generation, and hosting capacity. Although these problems are closely related to planning rather than operation, they share most of the properties of the OPF, and hence it is possible to obtain convex and mixed-integer convex approximations.

Mathematical Programming for Power Systems Operation: From Theory to Applications in Python. First Edition. Alejandro Garcés.
© 2022 by The Institute of Electrical and Electronics Engineers, Inc. Published 2022 by John Wiley & Sons, Inc.

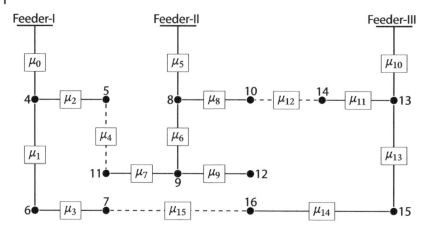

Figure 11.1 Three-feeder test system for power system reconfiguration [84].

11.2 Primary feeder reconfiguration

Power distribution networks have tie/sectionalizing switches μ_k that allow transferring load from one feeder to other, as depicted in Figure 11.1. This action may be performed automatically from the control center. However, we require an optimization algorithm that guides the process in order to minimize loss and improve efficiency. This algorithm is known as primary feeder reconfiguration [84].

In simple terms, the algorithm determines each switching state (on/off) that minimizes power loss. However, the problem is complex for three main reasons: first, the model must include power flow equations that are non-convex, as discussed in Chapter 10; second, the switches along the feeder impose binary constraints into the model; and third, it is required to impose constraints that guarantee that each primary feeder is radial and connected. We address each of these problems below.

We must represent the grid as an oriented graph $\mathcal{G} = \{\mathcal{N}, \mathcal{E}\}$, where $\mathcal{N}$ is the set of nodes and $\mathcal{E} \subseteq \mathcal{N} \times \mathcal{N}$ is the set of edges (connected or disconnected). Each node has associated a voltage v_k and a value of active and reactive power, p_k and q_k, respectively. Besides, each edge has an admittance y_{km} and a binary variable μ_{km} that represents the corresponding switch state. All substations are represented by the same node, marked as 0 and with voltage $v_0 = 1\angle 0$. Circuit relations are represented by the incidence matrix A, as presented below:

$$I_{\mathcal{N}} = AI_{\mathcal{E}} \tag{11.1}$$

$$V_{\mathcal{E}} = A^{\mathsf{T}} V_{\mathcal{N}} \tag{11.2}$$

where $I_{\mathcal{N}}$ and $V_{\mathcal{N}}$ are the vectors of nodal current and voltage, and $I_{\mathcal{E}}, V_{\mathcal{E}}$ are the vectors of branch current and voltage, respectively. The Ohm's law in each edge and the power balance in each node are also included into the model as follows:

$$i_{km} = \mu_{km} y_{km} v_{km}, \quad \forall km \in \mathcal{E} \tag{11.3}$$

$$i_k = (s_k/v_k)^*, \quad \forall k \in \mathcal{N} \tag{11.4}$$

$$\mu_{km} \in \{0, 1\} \tag{11.5}$$

these equations constitute the main binary and non-linear/non-convex constraints of the problem. Below, we propose a linear approximation for these constraints.

First, we define an auxiliary complex variable j_{km} for the current in each edge of the graph, regardless of whether the edge is connected or not. This current, given in 11.6, lacks physical meaning if the edge is disconnected and is used only as auxiliary variable.

$$j_{km} = y_{km} v_{km} \tag{11.6}$$

Then, the bi-linear equation related to current in each edge is replaced by a linear equivalent as explained in Chapter 4:

$$-\delta_I^{\text{real}} \mu_{km} \leq i_{km}^{\text{real}} \leq \delta_I^{\text{real}} \mu_{km} \tag{11.7}$$

$$-\delta_I^{\text{imag}} \mu_{km} \leq i_{km}^{\text{imag}} \leq \delta_I^{\text{imag}} \mu_{km} \tag{11.8}$$

$$j_{km}^{\text{real}} - (1 - \mu_{km})\delta_I^{\text{real}} \leq i_{km}^{\text{real}} \leq j_{km}^{\text{real}} + (1 - \mu_{km})\delta_I^{\text{real}} \tag{11.9}$$

$$j_{km}^{\text{imag}} - (1 - \mu_{km})\delta_I^{\text{imag}} \leq i_{km}^{\text{real}} \leq j_{km}^{\text{imag}} + (1 - \mu_{km})\delta_I^{\text{imag}} \tag{11.10}$$

where δ_I represents the maximum deviation of the current in each branch. Next, the power balance in each node is linearized using a complex linearization around $v_k = 1\angle 0^o$, as presented below[1]:

$$i_k = s_k^*(2 - v_k^*) \tag{11.11}$$

At this point, the model is a mixed-integer linear. However, it is required to impose a radiality constraint; otherwise, the model would connect all switches. A meshed grid is more efficient than a radial grid. However, radiality is required in classic power distribution networks because the protections are calibrated for such configuration. We use the radiality constraints proposed in [85], which are based on two key observations from graph theory: first, a radial grid (i.e, a tree)

---

1 See Appendix B for details about complex linearizations.

has $|\mathcal{E}| - 1$ node and second, the graph must be connected. The first observation can be imposed in the model as the following affine constraint:

$$\sum_{km \in \mathcal{E}} \mu_{km} = n - 1 \tag{11.12}$$

where n is the number of nodes. In this way, we ensure there are only $n - 1$ switches connected in the grid. The second condition can be imposed by noticing that the Laplacian matrix associated to the graph must be diagonally dominant. The Laplacian matrix W is defined as follows:

$$W = A \operatorname{diag}(\mu) A^{\mathsf{T}} \tag{11.13}$$

where $\operatorname{diag}(\mu)$ is a diagonal matrix of size $|\mathcal{E}| \times |\mathcal{E}|$. We could impose a constraint such that W is positive semidefinite in which case, we would obtain a semidefinite programming problem. However, it is straightforward to impose a simple linear constraint related to the diagonal-dominant characteristic of the Laplacian matrix, namely:

$$w_{kk} \geq \sum_{m} w_{km}, \ \forall k \tag{11.14}$$

Collecting all the aforementioned approximations, we obtain a mixed-integer linear programming model for the primary feeder reconfiguration. Let us see the use of the model by a simple example.

Example 11.1. Let us solve the power system reconfiguration problem in the classic three-feeder test system proposed by Civanlar in [84]. For the sake of completeness, the parameters of these feeders are presented in Table 11.1. We store all the parameters in a graph using the module `networkx` (see Appendix A). The inputs of the model are the matrix of admittance $Y_{\mathcal{E}}$, the incidence matrix, and the vector of nodal powers. The corresponding code for minimizing active power loss is presented below:

```
Vnode_real = cvx.Variable(num_nodes)
Vnode_imag = cvx.Variable(num_nodes)
Inode_real = cvx.Variable(num_nodes)
Inode_imag = cvx.Variable(num_nodes)
Vedge_real = cvx.Variable(num_edges)
Vedge_imag = cvx.Variable(num_edges)
Iedge_real = cvx.Variable(num_edges)
Iedge_imag = cvx.Variable(num_edges)
Jedge_real = cvx.Variable(num_edges)
Jedge_imag = cvx.Variable(num_edges)
W = cvx.Variable((num_nodes,num_nodes))
mu = cvx.Variable(num_edges, integer=True)
```

```
re = [mu >= 0, mu <= 1,
      Vnode_real[0] == 1,
      Vnode_imag[0] == 0,
      Vnode_real <= 1.2,
      Vnode_real >= 0.8,
      Vnode_imag <= 0.05,
      Vnode_imag >= -0.05,
      Vedge_real == A.T@Vnode_real,
      Vedge_imag == A.T@Vnode_imag,
      Jedge_real == Yedge_real@Vedge_real-Yedge_imag@Vedge_imag,
      Jedge_imag == Yedge_real@Vedge_imag+Yedge_imag@Vedge_real,
      Inode_real == A@Iedge_real,
      Inode_imag == A@Iedge_imag,
      W == A@cvx.diag(mu)@A.T,
      cvx.sum(mu) == num_nodes-1]
for k in range(1,num_nodes):
    re += [Inode_real[k]==S_real[k]*(2-Vnode_real[k])+
          S_imag[k]*(Vnode_imag[k])]
    re += [Inode_imag[k]==S_real[k]*(Vnode_imag[k])
          -S_imag[k]*(2-Vnode_real[k])]
    sm = 0
    for m in range(num_nodes):
        sm = sm + W[k,m]
    re += [sm >= 0]

for k in range(num_edges):
    re += [-mu[k]*deltaI_real[k] <= Iedge_real[k]]
    re += [-mu[k]*deltaI_imag[k] <= Iedge_imag[k]]
    re += [Iedge_real[k] <= mu[k]*deltaI_real[k]]
    re += [Iedge_imag[k] <= mu[k]*deltaI_imag[k]]
    re += [Iedge_real[k] <= Jedge_real[k]+(1-mu[k])
            *deltaI_real[k]]
    re += [Iedge_imag[k] <= Jedge_imag[k]+(1-mu[k])
            *deltaI_imag[k]]
    re += [Iedge_real[k] >= Jedge_real[k]-(1-mu[k])
            *deltaI_real[k]]
    re += [Iedge_imag[k] >= Jedge_imag[k]-(1-mu[k])
            *deltaI_imag[k]]

fo = cvx.Minimize(Inode_real[0])

Reconfiguration = cvx.Problem(fo,re)
```

Notice that minimizing power loss is equivalent to minimizing the power injected at the substation (i.e, i_0). The reader is invited to experiment with this code. 🐍

Table 11.1 Parameters of the three-feeder test system for power system reconfiguration [84].

From	To	r_{km}(pu)	x_{km}(pu)	p_k(pu)	q_k(pu)
SL	N4	0.0750	0.1000	0.02	0.02
N4	N5	0.0800	0.1100	0.03	0.00
N4	N6	0.0900	0.1800	0.02	0.00
N6	N7	0.0400	0.0400	0.02	0.01
SL	N8	0.1100	0.1100	0.04	0.03
N8	N9	0.0800	0.1100	0.05	0.02
N8	N10	0.1100	0.1100	0.01	0.01
N9	N11	0.1100	0.1100	0.01	−0.01
N9	N12	0.0800	0.1100	0.05	−0.02
SL	N13	0.1100	0.1100	0.01	0.01
N13	N14	0.0900	0.1200	0.01	−0.01
N13	N15	0.0800	0.1100	0.01	0.01
N15	N16	0.0400	0.0400	0.02	−0.01
N5	N11	0.0400	0.0400	0.00	0.00
N10	N14	0.0400	0.0400	0.00	0.00
N7	N16	0.0900	0.1200	0.00	0.00

11.3 Optimal placement of capacitors

Large primary feeders require reactive compensation (e.g., shunt capacitors) installed at appropriated locations in order to reduce power and energy losses and improve voltage profile [83]. An optimization model is required to define the size and place of each of these shunt capacitors, resulting in a problem that is closely related to the optimal power flow (OPF), previously studied in Chapter 10.

Shunt capacitors are represented as discrete injections of reactive power per unit. These capacitors can be fixed or switching capacitor banks located along the primary feeder. For the sake of simplicity, we address only the case of fixed capacitors.

The objective function consists in minimizing power loss and/or cost, whereas the result of the optimization is the size and placement of the shunt capacitors. It is not economically viable to place capacitors in all nodes and hence, the amount of reactive power must be limited. A basic optimization model is presented below:

$$\min f_{\text{objective}}$$

$$p_L \geq \text{real}\left(\sum_k \sum_m y_{km} v_k v_m\right)$$

$$v_0 = 1 + 0j$$

$$\delta \geq \|v_k - 1\|, \ \forall k \in \mathcal{N}$$

$$i_{km}^{\max} \geq \|y_{km}(v_k - v_m)\|, \ \forall km \in \mathcal{E} \qquad (11.15)$$

$$s_k^* - d_k^* - j\xi_k q^{\text{nom}} = \sum_m y_{km} v_k^* v_m, \ \forall k \in \mathcal{N}$$

$$\|s_k\| \leq s_k^{\max}, \ \forall k \in \mathcal{N}$$

$$\sum_k \xi_k \leq \xi^{\text{available}}$$

$$\xi_k \in \{0, 1\}, \ \forall k \in \mathcal{N}$$

Where q^{nom} is the nominal value of the shunt capacitors to be placed in the feeder and ξ_k is a binary variable that indicates the placement of one capacitor in node k; the amount of capacitors to be placed in the feeder is limited by $\xi^{\text{available}}$; the rest of the variables and constraints have the same interpretation as the OPF problem studied in Chapter 10.

Model Equation (11.15) has two sources of complexity: first, power flow equations are non-affine; and second, the problem is mixed-integer. The first issue can be addressed by using convex approximations, whereas the second issue is solved directly by CvxPy. Both linearization or conic approximations (i.e., SOC and SDP) can be used in this problem. A linearization is convenient since the model results in mixed-integer quadratic programming (MIQP), and these type of models are solvable in practice; the quadratic term comes from the power loss equation, which is convex. SOC and SDP approximations result in mixed-integer second-order and mixed-integer semidefinite programming problems, which are computationally more demanding.

We may be interested in minimizing operation costs in a planning period (e.g., one year). In that case, the objective function includes costs associated with the power and energy losses, as well as the cost of installation of shunt capacitors. The objective function is therefore given by Equation (11.16):

$$f_{\text{objective}} = f_{\text{loss}}(p_L, e_L) + f_{\text{installation}}(q) \qquad (11.16)$$

where e_L is the energy loss, and $f_{\text{loss}}, f_{\text{installation}}$ are functions of annual costs and installation, respectively. These functions are usually linear. The number of binary variables is the same in this case, but the feeder requires to be represented in a load curve in order to calculate both power and energy losses. Some

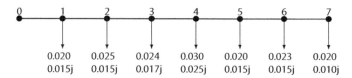

Figure 11.2 Radial distribution network $z_{km} = 0.01 + 0.005j$.

countries have different penalization costs for power and energy loss, and then the model requires to be adapted to each grid code.

Example 11.2. Let us consider the 8-nodes primary feeder shown in Figure 11.2; loads are depicted in the figure and impedance is $z_{km} = 0.01 + 0.005j$ for all line segments. We already calculated the matrix Y_{bus} and have a vector d of size n that stores the loads; up to 0.01pu reactive power compensation will be allowed. The optimization model is presented below:

```
n = 8
v = cvx.Variable(n,complex=True)        # voltages
W = cvx.Variable((n,n),complex=True)    # linearization
s0 = cvx.Variable(complex=True)         # power at slack
xi = cvx.Variable(n, boolean=True)      # shunt capacitors
pL = cvx.Variable()                     # power loss
s = n*[0]
s[0] = s0
M = Ybus@W
res = [pL >= cvx.quad_form(cvx.real(v),Ybus.real)+
       cvx.quad_form(cvx.imag(v),Ybus.real)]
res += [v[0]==1.0]
for k in range(n):
  res += [cvx.conj(s[k]+0.01j*xi[k]-d[k]) == M[k,k]]
  res += [cvx.abs(v[k]-1) <= 0.1]
  res += [xi[k]>= 0]
  res += [xi[k]<= 1]
  for m in range(n):
    res += [W[m,k] == cvx.conj(v[k])+v[m]-1]
res += [cvx.sum(xi) <= 1]
obj = cvx.Minimize(pL)
OPCAP = cvx.Problem(obj,res)
OPCAP.solve()
print('pL',pL.value)
print('shunt capacitors', np.round(xi.value))
```

In this case, we used a linear approximation of the power flow equations. The model places a capacitor at Node 7, resulting in a power loss of $p_L = 0.00102$. The reader is invited to experiment with this model; for example, relax the binary constraint (`boolean=False`) and compare the results.

Example 11.3. Power distribution networks may include active components such as D-STATCOMS (distribution static var compensators). These components are basically voltage source converters equipped with a suitable control that maintains a constant reactive power or nodal voltage. The model for optimal placement of capacitors can be easily extended to the optimal placement of D-STATCOMS. In that case, the term $\xi_k q^{nom}$ is replaced by a new variable $q_k = \xi_k q^{nom}$ in the power flow equations in Model Equation (11.15), and a new constraint is included as follows:

$$-\xi_k q^{nom} \leq q_k \leq \xi_k q^{nom} \tag{11.17}$$

Therefore, the optimization model returns not only the placement of the component but also the reactive power that must inject into the grid.

11.4 Optimal placement of distributed generation

Modern power distribution networks include a massive penetration of distributed generation, especially renewable sources such as photovoltaic and wind generation, motivated by a growing concern about global warming. The optimal placement of this distribution generation constitutes an optimization problem that can be efficiently solved by the convex approximations presented in the previous chapter. The problem is discrete, although a good approximation can be obtained if binary variables are relaxed [86].

The main objective is to minimize power loss p_L, although other objectives such as costs or reliability can be considered, subject to physical constraints similar to the OPF problem. A vector $\xi_k \in \{0, 1\}$ is defined for each node, where $\xi_k = 1$ if a distributed generator is placed in Node k. For the sake of simplicity, all generators are consider of the same capacity s^{nom}, resulting in the following optimization model:

$$\min p_L$$

$$p_L \geq \text{real}\left(\sum_k \sum_m y_{km} v_k v_m\right)$$

$$v_0 = 1 + 0j$$

$$\delta \geq \|v_k - 1\|, \ \forall k \in \mathcal{N}$$

$$i_{km}^{\max} \geq \|y_{km}(v_k - v_m)\|, \ \forall km \in \mathcal{E} \tag{11.18}$$

$$s_k^* - d_k^* = \sum_m y_{km} v_k^* v_m, \ \forall k \in \mathcal{N}$$

$$\|s_k\| \leq \xi_k s^{\text{nom}}, \ \forall k \in \mathcal{N}$$

$$\sum_m \xi_k \leq \xi^{\max}$$

$$\xi_k \in \{0, 1\}, \ \forall k \in \mathcal{N}$$

where $\xi^{\max}$ is the maximum number of distributed generators to be placed in the system. This is a mixed-integer convex programming problem that can be efficiently solved using mixed-integer methods such as the Branch and Bound method. The model returns not only the placement but also the sizing of distributed generators. The costs of these generators can be also included into the objective function, in any case, the model remains (mixed-integer) convex .

Example 11.4. We are interested in placing two distributed generators of $s^{\text{nom}} = 0.02\text{pu}$ in the primary distribution feeder presented in Figure 11.2, with the objective to minimize power loss. The code in Python for solving this problem is presented below (we assume parameters of the grid are stored in a graph G):

```python
n = G.number_of_nodes()
d = np.array([G.nodes[k]['d'] for k in G.nodes])

nt = 2
v = cvx.Variable(n, complex=True)
W = cvx.Variable((n,n), complex=True)
s = cvx.Variable(n, complex=True)
xi = cvx.Variable(n, nonneg=True)
pL = cvx.Variable()
M = Ybus@W
res = [pL >= cvx.quad_form(cvx.real(v), Ybus.real) +
       cvx.quad_form(cvx.imag(v), Ybus.real)]
res += [v[0]==1.0]
for k in range(n):
    res += [cvx.conj(s[k]-d[k]) == M[k,k]]
    res += [cvx.abs(v[k]-1) <= 0.05]
```

```
    for m in range(n):
        res += [W[m,k] == cvx.conj(v[k])+v[m]-1]
    for k in range(1,n):  # except the slack
      res += [cvx.abs(s[k]) <= 0.02*xi[k]]
      res += [xi[k]>=0, xi[k]<=1]
    res += [cvx.sum(xi) <= nt]

    obj = cvx.Minimize(pL)
    HOSTCAP = cvx.Problem(obj,res)
    HOSTCAP.solve()
    print(HOSTCAP.status,obj.value)
    print(np.round(xi.value,3))
```

Although the model is binary, we use a continuous relaxation that returns a final solution of $\xi_6 = \xi_7 = 1$. Notice that most of the code is similar to the OPF problem for a linear approximation. The reader is invited to test a SOC approximation for the problem. 🐍

11.5 Hosting capacity of solar energy

High penetration of renewable resources, especially solar photovoltaic, could create overvoltages along with primary feeders [87]. Therefore, it is necessary to define the amount of solar energy that can be hosted on a power distribution network without adversely impacting safety, power quality, reliability, or other operational features [1].

The hosting capacity model is similar to the OPF, however, in this case the capacity of each distributed generator is also a variable. The objective is to determine the maximum amount of power that can be generated in each node without jeopardizing the normal operation of the system. The model is presented below using a complex linearization of the power flow equations:

$$\max f_{\text{objective}}$$

$$v_0 = 1 + 0j$$

$$\delta \geq \|v_k - 1\|, \; \forall k \in \mathcal{N}$$

$$i_{km}^{\max} \geq \|y_{km}(v_k - v_m)\|, \; \forall km \in \mathcal{E} \tag{11.19}$$

$$s_k^* - d_k^* = \sum_m y_{km} v_k^* v_m, \; \forall k \in \mathcal{N}$$

$$(1/\rho)h_k \geq \|s_k\|, \; \forall k \in \mathcal{N} - \{0\}$$

$$h_k \geq \text{real}(s_k), \; \forall k \in \mathcal{N} - \{0\}$$

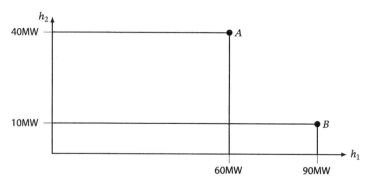

Figure 11.3 Example of two configurations of distributed generation.

In this model, h_k represents the maximum active power that can host each Node k; the objective function measures the maximum amount of power that can host the system, subject to the power flow equations and limits of voltage and current flow; ρ represents the minimum power factor in each power electronic converter.

There are different metrics to measure the hosting capacity, for instance we may be interested in maximizing the total distributed generation, in that case the objective function is given by Equation (11.20),

$$f_{\text{objective}} = \sum_k h_k \tag{11.20}$$

However, it may be the case that most of the distribution generation concentrates in a single node. To solve this problem, a metric based on the hypervolume of the new distribution generation is proposed, namely:

$$\max \prod_k h_k \tag{11.21}$$

where $\prod$ represents the product among the active power generated in the grid. In order to understand the logic behind Equation Equation (11.21) consider a system with two distributed generators h_1, h_2 with two possible configurations shown in Figure 11.3. Both configurations host the same amount of distributed generation, i.e., host(A) = 60MW + 40MW = 100MW and host(B) = 90MW + 10MW = 100MW; however, B concentrates most of the power in one node whereas A distribute the power more equitably; this can be measured by the area(A) = 2400, that is greater than area(B) = 2000. In a system with three generators, we can calculate the volume vol = $h_1 h_2 h_3$ instead of the area, and in the general case we can calculate the hypervolume given by Equation (11.21).

Equation Equation (11.21) can be transformed using a logarithmic function, as follows:

$$\ln\left(\prod_k h_k\right) = \sum_k \ln(h_k) \tag{11.22}$$

notice that ln is a monotone-concave function, and hence, we can define the following convex objective function:

$$\min \sum_k -\ln(h_k) \tag{11.23}$$

subject to the same constraints of Model Equation (11.19). In the following example, we compare objectives Equation (11.20) and Equation (11.23):

Example 11.5. Let us define the hosting capacity of the 8-nodes radial distribution network presented in Figure 11.2; we already defined a graph G with all the parameters of the grid. The code in Python for Model Equation (11.23) is presented below:

```python
n = G.number_of_nodes()
d = np.array([G.nodes[k]['d'] for k in G.nodes])

v = cvx.Variable(n, complex=True)
W = cvx.Variable((n,n), complex=True)
s = cvx.Variable(n, complex=True)
h = cvx.Variable(n)
M = Ybus@W
res = [v[0]==1.0]
for k in range(n):
    res += [cvx.conj(s[k]-d[k]) == M[k,k]]
    res += [cvx.abs(v[k]-1) <= 0.05]
    for m in range(n):
        res += [W[m,k] == cvx.conj(v[k])+v[m]-1]

htotal = 0
for k in range(1,n):   # except the slack
    htotal = htotal + h[k]
    res += [cvx.abs(s[k]) <= 1.2*h[k]]
    res += [cvx.real(s[k]) >= h[k]]
obj = cvx.Maximize(htotal)
HOSTCAP = cvx.Problem(obj,res)
HOSTCAP.solve()
print(HOSTCAP.status,obj.value)
print('hosting:',np.round(h.value,3))
```

After executed this code, a total power of 4.63pu is placed along the feeder with $h = (0, 4.54, 0.03, 0.02, 0.01, 0.01, 0.01, 0.01)^{\mathsf{T}}$; notice that most of the new generation is concentrated in a single node. However, in the case of the maximum hypervolume, the hosting capacity vector is $h = (0, 0.73, 0.37, 0.24, 0.18, 0.15, 0.12, 0.10)^\top$. The maximum hypervolume approach gives a better distribution of the newly distributed generation. 🐍

11.6 Harmonics and reactive power compensation

Power distribution systems may include non-linear loads, such as diode rectifiers and/or saturated magnetic devices, that introduce harmonic currents to the system. These harmonics, which are represented as currents at a multiple of the fundamental frequency, create power quality problems in the grid; so that they require to be reduced or, if possible, eliminated. One simple and efficient way to reduce these harmonics is by means of active filters as depicted in Figure 11.4. An active filter is a power electronic converter, usually a pulse-width modulated voltage-source converter, that injects currents that compensate for both reactive power and harmonic content. Most renewable resources, such as solar photovoltaics and wind energy, are integrated through these types of power electronic devices; hence, the compensation action may be performed by these devices.

A power electronic converter is able to control the output currents by using techniques such as carrier-based modulation, space vector modulation, or hysteresis control[2]. The reference of these currents is defined by a compensation theory. There are different compensation theories, as well as the definition of reactive power under harmonic distortion. In this section, we present a simple compensation theory based on mathematical optimization.

Indeed, the value of the current injected by the active filter, can be obtained by a simple optimization model. Let us consider a three-phase system with load currents i_A, i_B, and i_C; line-to-neutral voltages are given by v_A, v_B, and v_C; the currents injected by the active filter are u_A, u_B, and u_C. Therefore, the objective is to minimize the root mean square of the current line current, in one period; subject to power balance, as presented below:

$$\min \frac{1}{T} \int_{t}^{t+T} \left(\sum_{k \in \Phi} (i_k - u_k)^2 \right) dt'$$

$$\frac{1}{T} \int_{t}^{t+T} \left(\sum_{k \in \Phi} v_k u_k \right) dt' = 0 \qquad (11.24)$$

---

2 See [88] for more details about modulation and control of power electronic converters, for the integration of renewable energies.

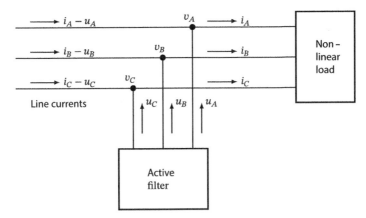

Figure 11.4 Schematic representation of the voltages and currents involved in the model for reactive power compensation and harmonic filtering.

where $T = 1/(2\pi f)$ and $\Phi = \{A, B, C\}$; notice that, minimizing line currents entails optimization in the power losses of the system. The constraint indicates that the instantaneous power delivered by the active filter is always zero in one period. This implies that, in average, the active filter does not deliver power to the grid.

This model is designed for real-time control, meaning that all variables are time-dependent. The optimization model is solved using Lagrange multipliers; therefore, the lagrangian function is calculated, namely:

$$\mathcal{L}(u, \lambda) = \frac{1}{T} \int_{t}^{t+T} \left(\sum_{k \in \Phi} (i_k - u_k)^2 \right) dt' + \frac{\lambda}{T} \int_{t}^{t+T} \left(\sum_{k \in \Phi} v_k u_k \right) dt' \qquad (11.25)$$

The first optimality conditions imply that the instantaneous derivative of L with respect to three-phase currents is equal to zero,

$$\frac{1}{T} \int_{t}^{t+T} (-2(i_k - u_k) + \lambda v_k) \, dt' = 0 \qquad (11.26)$$

For the integral to be zero, it is required that its integrand is also zero. Therefore, we have the following expresion:

$$-2(i_k - u_k) + \lambda v_k = 0 \qquad (11.27)$$

Therefore, the current in each phase is given by the equation presented below:

$$u_k = i_k - \frac{\lambda v_k}{2} \tag{11.28}$$

Let us multiply Equation (11.27) by v_k and add in the three phase, to obtain the following expression:

$$\sum_{k \in \Phi} -2i_k v_k + 2v_k u_k) + \lambda v_k^2 = 0 \tag{11.29}$$

Now, we integrate this expression and use the fact that the active filter does not deliver power in a period. Therefore, the following expression is obtained:

$$\frac{1}{T} \int_t^{t+T} \left(\sum_{k \in \Phi} (-2i_k v_k + v_k u_k + \lambda v_k^2) \right) dt' = 0 \tag{11.30}$$

$$-2\overline{p} + \lambda \overline{v}^2 = 0 \tag{11.31}$$

where $\overline{p}$ is the average power, given by the equation presented below:

$$\overline{p} = \frac{1}{T} \int_t^{t+T} \left(\sum_{k \in \Phi} v_k i_k \right) dt' \tag{11.32}$$

and $\overline{v}^2$ is the three-phase square voltage, given by the following expression:

$$\overline{v}^2 = \frac{1}{T} \int_t^{t+T} \left(\sum_{k \in \Phi} v_k^2 \right) dt' \tag{11.33}$$

$$= v_{A(\text{rms})}^2 + v_{B(\text{rms})}^2 + v_{C(\text{rms})}^2 \tag{11.34}$$

Finally, replacing Equation (11.31) into Equation (11.28), the compensation current is obtained.

$$u_k = i_k - \frac{\overline{p}}{\overline{v}^2} v_k \tag{11.35}$$

This simple expression defines the optimal current that the active filter must inject in order to reduce line currents and power loss. One aspect that is missing in this description is the effect on the power factor of the proposed approach. The example presented below shows this effect in practice.

Example 11.6. Let us consider a non-linear load with the fifth harmonic. To analyze this load, we define first a function that generates three-phase variables at the positive sequence, as presented in the code below:

Figure 11.5 Three-phase currents for a non-linear load with 5th harmonic.

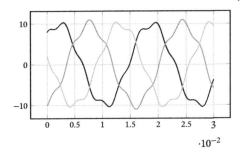

```
import NumPy as np
import matplotlib.pyplot as plt

t = np.linspace(0,2/60,100)
w = 2*np.pi*60
def three_phase(w,m,ph):
  xA = m*np.cos(w*t+ph)
  xB = m*np.cos(w*t+ph-2*np.pi/3)
  xC = m*np.cos(w*t+ph+2*np.pi/3)
  return xA, xB, xC
```

This function, receives the nominal frequency w, a magnitude m, and a phase ph. Nesx, the function is used to generate three-phase voltages and currents, as follows:

```
vA,vB,vC = three_phase(w,170,0)    # voltage
iA,iB,iC = three_phase(w,10,-0.8)  # fundamental current
hA,hB,hC = three_phase(5*w,1,0.1)  # harmonic current

iA = iA+hA
iB = iB+hB
iC = iC+hC
```

Figure 11.5 shows three-phase currents.

The compensation current is calculated using Equation (11.35) as presented below:

```
pm = vA*iA+ vB*iB + vC*iC
vm = (vA**2).mean() + (vB**2).mean() + (vC**2).mean()
uA = iA - pm.mean()/vm*(vA)
uB = iB - pm.mean()/vm*(vB)
uC = iC - pm.mean()/vm*(vC)
```

This compensation not only reduces the harmonic contents of the line currents but also achieves a unity power factor. The reader is invited to plot currents $i_k - u_k$ and analyze the results.

11.7 Further readings

All methods presented in this chapter may be extended to the case of three-phase unbalanced power grids. In that case, there is required an efficient way to store the parameters of the system. However, the main ideas are the same. The interested reader can be referred to [89].

Heuristic algorithms have also been proposed to solve the optimal capacitor placement as well as the optimal placement of distributed generation. See for example [83] and [90] for the capacitor problem and [91] for the distributed generation problem. Switched capacitors under unbalanced representation can be included in the problem as given in [92]. Typically, these algorithms use a master-slave strategy where a master problem chooses the placement and size of the component, and the slave algorithm solves a power flow to obtain the continuous variables. However, the random nature of the solutions obtained by metaheuristics makes them unsuitable for real applications. In addition, the abuse of biological and social metaphors tends to hide the mathematical and physical structure of the original problem [93].

In the case of the hosting capacity problem, it can be solved using Monte Carlo simulation [94]. This method generates a high number of scenarios where a power flow analysis is performed [95]. The elevated number of scenarios makes the method cumbersome for everyday operation. Risk assessment tools have also been proposed to solve the problem [96]. However, it also requires the generation of multiple scenarios, just as in the case of Montecarlo methods. Heuristic and metaheuristic methods have also been proposed for solving these types of models [97]. However, a convex approximation, like the one presented in this chapter, is sufficient to solve the problem.

The theory presented here for compensation of reactive power and active filtering is, perhaps, the most simple approach to solve this problem. However, there are several methodologies and theories which may be complete and rigorous. These theories can include the effect of the neutral current and other compensation objectives, as was demonstrated in [98]. Other compensation theories can be found in [99] and [88], and in seminal papers such as [100] and [101].

11.8 Exercises

1. Solve the problem presented in Example 11.2 using a SOC formulation for the power flow equations. Compare the results.
2. Solve the problem of optimal placement of D-STATCOMs in the system presented in Example 11.2. Use the same parameters of costs and q^{nom}.

3. Find the optimal capacitor placement for the 34-bus test system presented in Table 10.1 (Chapter 10); T1 represents the load factor for operation between 0h:8h, T2 for 8h:16h, and T3 for 16h:24h. Try different objective functions, for example power loss, energy loss, and/or costs.

4. Solve the problem presented in Example 11.4 considering binary variables ξ_k and $s^{nom} = 0.3$pu.

5. Solve the problem presented in Example 11.4 using an SDP approximation for the power flow equations. Compare the results.

6. Solve the primary feeder reconfiguration for the test system presented in Table 11.2.

7. Solve the problem presented in Example 11.5 using a SOC approximation for the power flow equations.

8. Consider the problem presented in Example 11.5 but now, a loop is created connecting nodes 2 and 5. Compare the results.

9. Determine the hosting capacity for the power distribution network presented in Table 10.1 (Chapter 10).

10. The concept of hosting capacity can be extended to dc distribution grids. Formulate and solve the problem in the 21-nodes dc-distribution system presented in Table 10.2.

Table 11.2   IEEE 33 nodes test distribution network [102].

From	To	r_{km}(pu)	x_{km}(pu)	p_k(pu)	q_k(pu)
0	1	0.000575259	0.000297612	0.100	0.060
1	2	0.003075952	0.001566676	0.090	0.040
2	3	0.002283567	0.001162997	0.120	0.080
3	4	0.002377779	0.001211039	0.060	0.030
4	5	0.005109948	0.004411152	0.060	0.020
5	6	0.001167988	0.003860850	0.200	0.100
6	7	0.010677857	0.007706101	0.200	0.100
7	8	0.006426430	0.004617047	0.060	0.020
8	9	0.006488823	0.004617047	0.060	0.020
9	10	0.001226637	0.000405551	0.045	0.030
10	11	0.002335976	0.000772420	0.060	0.035
11	12	0.009159223	0.007206337	0.060	0.035
12	13	0.003379179	0.004447963	0.120	0.080
13	14	0.003687398	0.003281847	0.060	0.010
14	15	0.004656354	0.003400393	0.060	0.020
15	16	0.008042397	0.010737754	0.060	0.020
16	17	0.004567133	0.003581331	0.090	0.040
1	18	0.001023237	0.000976443	0.090	0.040
18	19	0.009385084	0.008456683	0.090	0.040
19	20	0.002554974	0.002984859	0.090	0.040
20	21	0.004423006	0.005848052	0.090	0.040
2	22	0.002815151	0.001923562	0.090	0.050
22	23	0.005602849	0.004424254	0.420	0.200
23	24	0.005590371	0.004374340	0.420	0.200
5	25	0.001266568	0.000645139	0.060	0.025
25	26	0.001773196	0.000902820	0.060	0.025
26	27	0.006607369	0.005825590	0.060	0.020
27	28	0.005017607	0.004371221	0.120	0.070
28	29	0.003166421	0.001612847	0.200	0.600
29	30	0.006079528	0.006008401	0.150	0.070
30	31	0.001937288	0.002257986	0.210	0.100
31	32	0.002127585	0.003308052	0.060	0.040
7	20	0.012478500	0.012478500	0	0
8	14	0.012478500	0.012478500	0	0
11	21	0.012478500	0.012478500	0	0
17	32	0.003119600	0.003119600	0	0
24	28	0.003119600	0.003119600	0	0

12

State estimation and grid identification

Learning outcomes

By the end of this chapter, the student will be able to:

- Solve basic state estimation problems using the gradient method.
- Identify the Y_{bus} from measurements of voltage and current.
- Solve optimization problems including norms in the objective function.

12.1 Measurement units

Synchrophasor or phasor measurement units (PMUs) are devices that allow measuring voltage and current in magnitude and angle via global positioning system (GPS) synchronization, i.e., the time-stamp given by the GPS is used to synchronize measures and obtain exact values of nodal angles. PMUs are common in modern power systems and constitute the primary tool to improve observability. However, measures alone are not enough to have an accurate picture of the state of the grid. Therefore, a state estimation algorithm must filter redundant data and compensate spurious measurements [103]. This algorithm is integrated into the supervisory control and data acquisition system (SCADA), which requires to be precise, exact, and highly efficient to operate in real-time.

The use of PMUs also allows estimating the Y_{bus} and even the grid's topology in a model known as grid identification or inverse power flow [104]. Both the state estimation and the grid identification are studied in this chapter under the assumption there are PMUs in all nodes. In order to keep our philosophy of toy-models, our presentation is based on the module CvxPy. However, practical

Mathematical Programming for Power Systems Operation: From Theory to Applications in Python. First Edition. Alejandro Garcés.
© 2022 by The Institute of Electrical and Electronics Engineers, Inc. Published 2022 by John Wiley & Sons, Inc.

implementations use tailored algorithms based on the gradient method, which are faster and more efficient.

12.2 State estimation

Measurements instruments such as voltmeters and ammeters are never perfect but have an intrinsic measurement error. Therefore, the actual state of a system is always unknown, although it can be estimated using available measures and an optimization method known as state estimation.

To understand the problem, let us consider a simple circuit made up of a voltage source and a resistor. Let us suppose the resistance is $5.0\,\Omega$, and a voltmeter connected in parallel to this resistor gives a voltage of 9.6 V, whereas an ammeter connected in series gives a current of 1.83 A. We know that the circuit must comply with Ohm's law, but $1.83\,\text{A} \times 5\Omega = 9.15\text{V} \neq 9.6\text{V}$, which measurement should we trust? The voltmeter or the ammeter? It is not a large discrepancy, but this type of error can spread in a system with thousands of measurements. We require a systematic approach.

Consider a power system with different type of real-time measurement instruments as well as pseudo-measurements (i.e forecasts or historical data). then, a non-linear measurement model may be defined as (12.1):

$$z = h(x) + e \tag{12.1}$$

where $z \in \mathbb{R}^m$ is the vector of measurement (and pseudo-measurements), $x \in \mathbb{R}^n$ is the true state vector, $h : \mathbb{R}^n \to \mathbb{R}^m$ relates measurements and states, and $e \in \mathbb{R}^m$ is the measurement error. Dimension of z is higher than dimension of x ($m > n$) in order to obtain an over determined system of non-linear equations[1]. We only know z and h, thus, our objective is to find an estimate for x such that the estimation error is minimized. Therefore, the problem can be represented as weighted least squares model:

$$\min \frac{1}{2} (z - h(x))^\top W (z - h(x)) \tag{12.2}$$

where W is a diagonal matrix that represents the weight associated to each measurement. We assume each measurement instrument is independent with zero mean error and variance σ_i^2, therefore, $W = \text{diag}(1/\sigma_i^2)$. The problem may be complemented with other equality and inequality constraints that represent operating limits and unobservable parts of the network; and for the sake of

---

1 In our naive example, we have one state $x = i$, two equation $h_1 = Ri, h_2 = i$, and two measurements $z_1 = v, z_2 = i$.

simplicity, we focus on the unrestricted case, the reader who wishes to delve into the subject can refer to [103].

In the classic formulation of the problem, the state variables x are the nodal voltage (magnitude and angle), whereas the measurements z include voltage magnitudes, active and reactive power flows, active and reactive power injections, and current magnitudes, among others. Modern state estimation models include phasor measurement units which allow obtaining the angles as real-time and synchronized measurements.

In general, Model (12.2) is non-convex since h is made up of non-linear relations between states and measurements. In those cases, the problem is solved using Newton-based methods without guarantee of finding the global optimum. The general case includes inequality constraints, and hence it is solved using interior-point methods, again, without a theoretical guarantee of convergence or optimality. However, we are interested in linear measurement models that make the problem convex and solvable in CvxPy. On the positive side, this approach is close to reality to the extent that modern systems rely on PMUs, which generate linear relationships between states and measurements. Moreover, we seek for toy-models that allow us to understand the problem and solve it using the paradigm of disciplined convex optimization. On the negative side, a state estimation algorithm is mainly designed for real-time operation, and hence Python may not be the best option in practice since algorithms implemented in Python tend to be slower than their counterpart in compiled languages C or C++.

The most simple instance of the problem is the dc state estimation. In this case, the grid is represented by the linear equations or DC power flow; hence, the state of the system is given by the angle of nodal voltages θ. Our objective is to find a vector x_{th} that estimates θ using available measurements of power at each substation. Three type of measurements are available as shown in Figure 12.1, namely: nodal powers z_{p}, power flows departing from the node z_{pf1}, and power flows arriving to the node z_{pf2}. Nodal powers and power flows in the transmission lines are linearly related to the nodal powers as follows:

$$z_{\text{p}} = Bx_{\text{th}} + e_{\text{p}} \tag{12.3}$$

Figure 12.1 Power measurements at a given node i for dc state estimation.

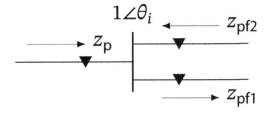

$$z_{pf1} = Hx_{th} + e_{pf1} \tag{12.4}$$

$$z_{pf2} = -Hx_{th} + e_{pf2} \tag{12.5}$$

where B is the Jacobian for the linear (dc) formulation of the power flow given by (12.12) as function of the nodal admittance matrix A and the branch admittance Y:

$$B = AYA^\mathsf{T} \tag{12.6}$$

and H is a matrix representation of the power flows, namely:

$$H = YA^\mathsf{T} \tag{12.7}$$

The error associated to each measurement has a variance $\sigma_p^2, \sigma_{pf1}^2$, and σ_{pf2}^2 which allow to define weight factors as follows:

$$W_p = \mathrm{diag}(1/\sigma_p^2) \tag{12.8}$$

$$W_{pf1} = \mathrm{diag}(1/\sigma_{pf1}^2) \tag{12.9}$$

$$W_{pf2} = \mathrm{diag}(1/\sigma_{pf2}^2) \tag{12.10}$$

These weight factors allow to formulate the following optimization model for estimating the angles of the system:

$$
\begin{aligned}
\min \ & \frac{1}{2}(z_p - Bx_{th})^\mathsf{T} W_p (z_p - Bx_{th}) \\
& + \frac{1}{2}(z_{pf1} - Hx_{th})^\mathsf{T} W_{pf1}(z_{pf1} - Hx_{th}) \\
& + \frac{1}{2}(z_{pf2} - Hx_{th})^\mathsf{T} W_{pf2}(z_{pf2} - Hx_{th})
\end{aligned}
\tag{12.11}
$$

The model may be complemented with additional constraint $x_{th}(0) = 0$ to ensure the slack node has an angle equal to zero. This model is evidently quadratic-convex.

Example 12.1. Figure 12.2 shows a network with three buses, two generators, and a load. All voltages have the same magnitude ($1pu$), but their angles (θ_i) are unknown. Power metering systems are placed to measure both nodal power and power flows. The state variables are θ_i, and the measurements can be related to the states via linear equations resulting in a convex problem.

The state of the system can be completely represented by the angles of the system $x_{th} = (\theta_0, \theta_1, \theta_2)$. In addition, there are three set of measurements, namely: $z_p = (p_0, p_1, p_2)^\mathsf{T}$ for nodal powers, $z_{pf1} = (p_{01}, p_{02}, p_{12})^\mathsf{T}$ for the power flows measured at the beginning of the lines, and $z_{pf2} = (p_{10}, p_{20}, p_{21})^\mathsf{T}$ for the power flows measured at the end of the lines. Each of these vectors

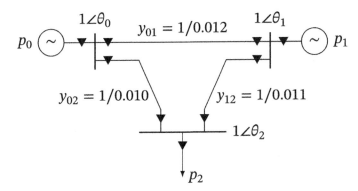

Figure 12.2 Three-bus system. ▼ represents points with power metering systems.

have a linear relation with the nodal angles as given in (12.3) to (12.5) with the following numerical values for matrices B and H:

$$B = \begin{pmatrix} 1/0.012 + 1/0.010 & -1/0.012 & -1/0.010 \\ -1/0.012 & 1/0.012 + 1/0.011 & -1/0.011 \\ -1/0.010 & -1/0.011 & 1/0.011 + 1/0.010 \end{pmatrix}$$

(12.12)

$$H = \begin{pmatrix} 1/0.012 & -1/0.012 & 0 \\ 1/0.010 & 0 & -1/0.010 \\ 0 & 1/0.011 & -1/0.011 \end{pmatrix}$$

(12.13)

Let us suppose nodal powers are $p = (0.8, 0.7, -1.5)^\mathsf{T}$, therefore, nodal angles $\theta = $ th and power flows pf can be calculated as follows:

```python
import networkx as nx
G = nx.DiGraph()
G.add_edges_from([(0,1),(0,2),(1,2)])
A = nx.incidence_matrix(G,oriented=True)
Y = 1/np.array([0.012,0.010,0.011])
B = A@np.diag(Y)@A.T
p = np.array([0.8,0.7,-1.5])
th = np.linalg.solve(B,p)
th = th - th[0] # the angle in the slack is 0
H = -np.diag(Y)@A.T
pf1 = H@th  # power flow at the beginning of the line
pf2 = -pf1  # power flow at the end of the line
```

This is the real state of the grid, however, we can only obtain measurements with normal distributed noise with zero mean and variance σ_p and σ_f for the nodal powers and the power flows, respectively. This effect can be considered as follows:

```
sigma_p = 0.01
sigma_f = 0.01
zp = p + np.random.randn(3)*sigma_p
zf1 = pf1 + np.random.randn(3)*sigma_f
zf2 = pf2 + np.random.randn(3)*sigma_f
```

With these measurements we have all the elements to formulate Model (12.11) as given below:

```
import cvxpy as cvx
Wp = (1/sigma_p**2)*np.identity(3)
Wf = (1/sigma_f**2)*np.identity(3)
x_th = cvx.Variable(3)
obj = cvx.Minimize(1/2*cvx.quad_form(zp-B@x_th,Wp)+
                   1/2*cvx.quad_form(zf1-H@x_th,Wf)+
                   1/2*cvx.quad_form(zf2+H@x_th,Wf))
res = [x_th[0] == 0]
WLS = cvx.Problem(obj,res)
WLS.solve(verbose=True)
```

We can know how accurate our model is by calculating the distance between θ (the real state) and x_{th} (the estimation).

```
print(np.linalg.norm(x_th.value-th)*100)
```

The reader is invited to experiment with the model by executing the code several times with different values of σ_p and σ_{pf}.

Example 12.2. Consider the power distribution grid depicted in Figure 10.1, Chapter 10. The Y_{bus} is calculated as Example 10.2 and the real state of the system is obtained by load flow algorithm given in Example 10.3, then, nodal values can be calculated as given below:

```
d = np.array([G.nodes[k]['d'] for k in G.nodes])
s = np.array([G.nodes[k]['smax'] for k in G.nodes])
Vn = LoadFlow(s[1:n],d[1:n])
In = Ybus@Vn
Sn = Vn*In.conj()
```

Let us suppose this is the real state of the system, however, we have only inaccurate measurements of voltages and currents; this can be represented in the model as measurements z_v, z_i with random noise, namely:

```
zv = Vn + np.random.randn(n)*0.001*np.exp(0.1j*np.random.
     randn(n))
zi = In + np.random.randn(n)*0.0001*np.exp(0.1j*np.random.
     randn(n))
```

Both voltages and currents are inaccurate so we cannot rely on one to calculate the other. However, we can generate the following measurement model in complex variable:

$$\begin{pmatrix} z_v \\ z_i \end{pmatrix} = \begin{pmatrix} 1_n \\ Y_{\text{bus}} \end{pmatrix} x_v + \begin{pmatrix} e_v \\ e_i \end{pmatrix} \tag{12.14}$$

where the state variables are represented by the vector x_v that estimates the nodal voltages and the measurements are $z = (z_v, z_i)^\mathsf{T} = (V_n + e_v, I_n + e_i)^\mathsf{T}$. The matrix 1_n is the identity of size n and the errors are given by $(e_v, e_i)^\mathsf{T}$. We assume both voltages and currents has the same accuracy and hence the weight matrix can be given as the identity. Therefore, Model (12.2) can be directly implemented in Python as follows:

```python
import cvxpy as cvx
xv = cvx.Variable(n, complex=True)
id = np.identity(n)
obj = cvx.Minimize(1/2*cvx.quad_form(zv-xv,id)+
                   1/2*cvx.quad_form(zi-Ybus*xv,id))
WLS = cvx.Problem(obj)
WLS.solve(verbose=True)
```

At first glance, it would seem illogical to calculate x_v given that we have a set of voltage measurements z_v, however, note that x_v has a smaller deviation from the true state of the system V_n (which we do not know in practice), thanks to the information provided by the other measurements. Let us calculate this deviation in percentage:

```python
print(np.linalg.norm(zv-Vn)*100)
print(np.linalg.norm(xv.value-Vn)*100)
```

After executing this code, the deviation of z_v with respect to V_n is around 0.4%, whereas the deviation of the estimation is around 0.03% (results can change from one execution to another due to the random error introduced in the code).

12.3 Topology identification

Power distribution networks are usually operated radially. However, along with primary feeders, there are tie and sectionalizing switches that allow changing the topology, transferring load from one feeder to another. Modern sectionalizing switches may be controlled centrally, but the switching effect requires to be checked. This observability problem for smart-distribution networks is known as topology identification.

On the other hand, the increasing growth of measurement technologies for power systems applications, such as smart meters and PMUs, results in improved controllability and observability. These aspects are essentials for the development of smart-grids at power and distribution levels. However, the actual implementation of these technologies in power distribution networks is limited by their costs. Therefore, early implementation of smart-distribution applications shall come with low-cost technologies that have limited measurement capability. Therefore, we require efficient algorithms for topology identification that guarantee real-time operation using low-cost measurement technologies [105].

Let us consider a power distribution network as the one depicted in Figure 12.1. We suppose the system is equipped with sectionalizing switches that modify the topology according to an optimization model that seeks loss reduction (See Section 11.2 Chapter 11 for the distribution feeder reconfiguration problem). The grid is also equipped with a non-contact line current sensor at specific points. A set of pseudo measurements of the nodal power is also considered in the problem.

The grid is represented as an oriented graph $\mathcal{G} = \{\mathcal{N}, \mathcal{E}\}$ with $\mathcal{N} = \{0, 1, \ldots, k, \ldots n\}$ the set of nodes and $\mathcal{E} \subseteq \mathcal{N} \times \mathcal{N}$ the set of edges. The following variables are considered in the model: the nodal voltage $V_{\mathcal{N}} = [v_k]$ estimated at node k; the current $I_{\mathcal{N}} = [i_k]$ estimated at node k; the edge current $I_{\mathcal{E}} = [i_{km}]$ estimated at branch km; and a binary variable μ_{km} that represents the switching status of edge km. Besides, the following inputs and parameters are considered: the pseudo-measurement of the model power s_k at each node k; a subset $\mathcal{M} \subset \mathcal{E}$ that represents the edges with current sensors; the measurement of these current sensors ξ_{km}; the admittance of each branch of the grid $Y_{\mathcal{E}} = [g_{km}]$; and the incidence matrix A of the graph included all the branches.

The optimization model consists of minimizing the error between measured and estimated variables, subject to power flow constraints as presented below:

$$\min \sum_{km \in \mathcal{M}} |i_{km} - \xi_{km}|$$

$$\mu_{km} \in \{0, 1\}$$

$$V_{\mathcal{N}} = A^{\mathsf{T}} V_{\mathcal{E}} \tag{12.15}$$

$$I_{\mathcal{N}} = A I_{\mathcal{E}}$$

$$I_{\mathcal{E}} = \mu Y_{\mathcal{E}} V_{\mathcal{E}}$$

$$i_k = (s_k / v_k)^*$$

This model is non-linear and mixed-integer. Hence, a mixed-integer linear approximation is developed as follows: first, the non-linear equation of the nodal power is approximated to a linear model as follows [89]:

$$i_k = s_k^*(2 - v_k^*) \tag{12.16}$$

Next, the product of binary and continuous variables is approximated as presented below[2]:

$$-\mu_{km}\delta_{km} \leq i_{km}^{\text{real}} \leq \mu_{km}\delta_{km}$$

$$-\mu_{km}\delta_{km} \leq i_{km}^{\text{imag}} \leq \mu_{km}\delta_{km}$$

$$\lambda_{km}^{\text{real}} - (1 - \mu_{km})\delta_{km} \leq i_{km}^{\text{real}} \leq \lambda_{km}^{\text{real}} + (1 - \mu_{km})\delta_{km} \tag{12.17}$$

$$\lambda_{km}^{\text{imag}} - (1 - \mu_{km})\delta_{km} \leq i_{km}^{\text{imag}} \leq \lambda_{km}^{\text{imag}} + (1 - \mu_{km})\delta_{km}$$

$$\lambda_{km} = y_{km}v_{km}$$

where λ_{km} is an auxiliary complex variable related to the current for each branch km and δ_{km} is the current capacity of each branch.

Example 12.3. The mixed-integer model for topology identification in power distribution is implemented in Python as follows:

```
Vnode_real = cvx.Variable(num_nodes)
Vnode_imag = cvx.Variable(num_nodes)
Inode_real = cvx.Variable(num_nodes)
Inode_imag = cvx.Variable(num_nodes)
Vedge_real = cvx.Variable(num_edges)
Vedge_imag = cvx.Variable(num_edges)
Iedge_real = cvx.Variable(num_edges)
Iedge_imag = cvx.Variable(num_edges)
Jedge_real = cvx.Variable(num_edges)
Jedge_imag = cvx.Variable(num_edges)
mu = cvx.Variable(num_edges, integer=True)
re = [mu >= 0, mu <= 1,
      Vnode_real[0] == 1,              Vnode_imag[0] == 0,
      Vedge_real == A.T@Vnode_real, Vedge_imag ==
                      A.T@Vnode_imag,
      Inode_real == A@Iedge_real,    Inode_imag ==
                      A@Iedge_imag,
      Jedge_real == Yedge_real@Vedge_real-Yedge_imag@
                      Vedge_imag,
      Jedge_imag == Yedge_real@Vedge_imag+Yedge_imag@
                      Vedge_real]
for k in range(1,num_nodes):
    re += [Inode_real[k]==S_real[k]*(2-Vnode_real[k])+
```

---

2 See Section 4.9 Chapter 4.

```
                    S_imag[k]*(Vnode_imag[k])]
        re += [Inode_imag[k]==S_real[k]*(Vnode_imag[k])-
                    S_imag[k]*(2-Vnode_real[k])]
for k in range(num_edges):
        re += [-mu[k]*deltaI_real[k] <= Iedge_real[k]]
        re += [-mu[k]*deltaI_imag[k] <= Iedge_imag[k]]
        re += [Iedge_real[k] <= mu[k]*deltaI_real[k]]
        re += [Iedge_imag[k] <= mu[k]*deltaI_imag[k]]
        re += [Iedge_real[k] <= Jedge_real[k]+(1-mu[k])
                    *deltaI_real[k]]
        re += [Iedge_imag[k] <= Jedge_imag[k]+(1-mu[k])
                    *deltaI_imag[k]]
        re += [Iedge_real[k] >= Jedge_real[k]-(1-mu[k])
                    *deltaI_real[k]]
        re += [Iedge_imag[k] >= Jedge_imag[k]-(1-mu[k])
                    *deltaI_imag[k]]
fo  = cvx.sum(Wmes@cvx.abs(Iedge_real-IE.real))
fo += cvx.sum(Wmes@cvx.abs(Iedge_imag-IE.imag))
Identification = cvx.Problem(cvx.Minimize(fo),re)
Identification.solve(verbose=True)
```

where the input are the measurements IE, the incidence matrix A, and the pseudo-measurements of power S_real and S_imag. The model was separated into real and imaginary parts in order to simplify its implementation.

12.4 Y_{bus} estimation

Estimating the Y_{bus} is a crucial problem in the operation of power systems, both at power and distribution levels. We are interested in generating an estimation Y closest to the real value Y_{bus}. Therefore, we require a way to measure the distance between Y and Y_{bus}, that is to say, we require a norm, as defined in Chapter 2, but in the space of the complex matrices; so, given a norm $\|\cdot\|$ in $\mathbb{R}^n$, we can generate a new norm in space $\mathbb{R}^{m \times n}$ as given in (12.18).

$$\|M\| = \sup_{\|x\|=1} \|Mx\| \tag{12.18}$$

As the norm of a vector, a matrix norm is a convex function that may be used as the objective function is convex optimization problems implemented in CvxPy.

Consider a power grid in which all nodes are equipped with PMUs that allow measuring both voltages and currents. Let us assume we do not know the exact values of the parameters of the transmission lines and the topology of

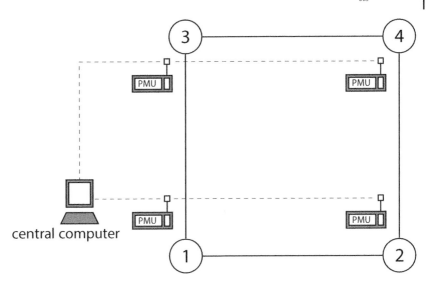

Figure 12.3 Four-node grid with PMUs in all nodes for Y_{bus} identification.

the grid, and therefore, an Y_{bus} estimation algorithm is required. A central computer receives and stores the information for different scenarios as depicted in Figure 12.3.

The basic estimation algorithm is defined as the following unconstrained optimization problem:

$$\min_{Y} \ \|YV - I\|^2 \tag{12.19}$$

where $\|\cdot\|$ is any matrix norm, V and I are matrices of size $n_n \times n_s$ with n_n the number of nodes and n_s the number of scenarios. Notice this optimization problem is defined in the set of the complex matrices and not in the set of vectors as in the optimization problems presented in previous chapters.

The model may be complemented with additional constraints related to the main features of the Y_{bus}. For instance, we know it is symmetric, meaning that

$$Y = Y^{\mathsf{T}} \tag{12.20}$$

In addition, we know that $G = \text{real}(Y)$ and $B = -\text{imag}(Y)$ are positive semidefinite, diagonally dominant and sparse. The topology of the graph defines entries that are already known and equal to zero. All these features can be added to the model in order to obtain a more accurate estimation.

Example 12.4. Let us estimate the Y_{bus} for the system shown in Figure 12.3. Our example is divided into three parts: first, we generate the exact model of the grid, in order to obtain the correct value of the Y_{bus}; then, a random set of measurement scenarios are generated with this matrix; and finally, the Y_{bus} is estimated by the proposed optimization model.

We use the module NetworkX for generating the Y_{bus} as presented below; parameters of the transmission lines are included in the code:

```python
import numpy as np
import networkx as nx
Grid = nx.DiGraph()
Grid.add_edges_from(((1,2),(2,3),(3,4),(4,1)))
Yp = 1/np.array([0.002+0.02j,0.001+0.03j,0.001+0.02j,
        0.001+0.04j])
A = nx.incidence_matrix(Grid,oriented=True)
Ybus = A@np.diag(Yp)@A.T
```

Now, we generate a set of random scenarios for the voltages around 1pu and next, these voltages are used to calculate nodal currents for each scenario. Finally, a random noise is added in order to emulate possible inaccuracies of the measure devices. The code in Python is presented below:

```python
nn = 4      # number of nodes
ns = 20     # number of scenarios
v = 0.9*np.ones((nn,ns))+0.2*np.random.random((nn,ns))
a = np.random.random((nn,ns))-0.5
Vbus = v*np.exp(a*1j)
Ibus = Ybus@Vbus + np.random.normal(0,0.1,(nn,ns))
Ibus = Ibus + 1j*np.random.normal(0,0.1,(nn,ns))
```

The optimization model consists on minimizing (12.19) subject to the constraint $Y = Y^T$ as follows:

```python
import cvxpy as cvx
Y = cvx.Variable((nn,nn), complex=True)
fo = cvx.Minimize(cvx.norm(Y@Vbus-Ibus))
re = [Y==Y.T]
Est = cvx.Problem(fo,re)
Est.solve()
print(np.linalg.norm(Y.value-Ybus))
```

The result of this model is different each time the script is executed due to the random scenarios. However, the order of magnitude of the error is the same according to the number of scenarios. The script was executed with a different number of scenarios, and the results are shown in Figure 12.4; it shows that a high number of scenarios does not necessarily increase the accuracy of the estimation.

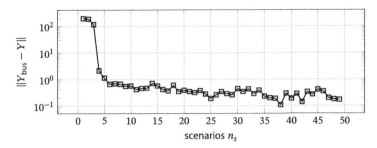

Figure 12.4 Estimation error vs number of scenarios.

Example 12.5. Consider now the previous example with semidefinite contraints as follows:

```
Y = cvx.Variable((nn,nn), complex=True)
fo = cvx.Minimize(cvx.norm(Y@Vbus-Ibus))
re = [Y==Y.T, cvx.real(Y) >> 0, -cvx.imag(Y)>>0]
Est2 = cvx.Problem(fo,re)
Est2.solve()
print(np.linalg.norm(Y.value-Ybus))
```

The reader is invited to execute the script for different number of scenarios n_s. In general, the positive semidefinite constraints does not improve the accuracy of the solution for a large set of scenarios (e.g $n_s = 20$), and instead, the time calculation is highly increased. However, for a small data set (for instance $n_s = 3$), the semidefinite constraints highly improve the results.

Example 12.6. We can show that the following conditions hold in the grid of Example 12.4:

$$g_{kk} \geq \sum_{m \neq k} -g_{km}$$

$$-b_{kk} \geq \sum_{m \neq k} b_{km} \tag{12.21}$$

This conditions are general, easy to implement, and less time-consuming that semidefinite constraints, for problems with few measurement scenarios. A script with this constraints is presented below:

```
Y = cvx.Variable((nn,nn), complex=True)
fo = cvx.Minimize(cvx.norm(Y@Vbus-Ibus))
re = [Y==Y.T]
re += [cvx.real(Y[0,0]) >= -cvx.real(Y[0,1]+Y[0,2]+Y[0,3])]
re += [cvx.real(Y[1,1]) >= -cvx.real(Y[1,0]+Y[1,2]+Y[1,3])]
re += [cvx.real(Y[2,2]) >= -cvx.real(Y[2,0]+Y[2,1]+Y[2,3])]
re += [cvx.real(Y[3,3]) >= -cvx.real(Y[3,0]+Y[3,1]+Y[3,2])]
re += [-cvx.imag(Y[0,0]) >= cvx.imag(Y[0,1]+Y[0,2]+Y[0,3])]
```

```
re += [-cvx.imag(Y[1,1]) >= cvx.imag(Y[1,0]+Y[1,2]+Y[1,3])]
re += [-cvx.imag(Y[2,2]) >= cvx.imag(Y[2,0]+Y[2,1]+Y[2,3])]
re += [-cvx.imag(Y[3,3]) >= cvx.imag(Y[3,0]+Y[3,1]+Y[3,2])]
Est2 = cvx.Problem(fo,re)
Est2.solve()
print(np.linalg.norm(Y.value-Ybus))
```

The student is invited to compare results with the previous examples for $n_s = 3$.

12.5 Load model estimation

The loads in distribution systems depend on the voltage, therefore, they are usually represented by quadratic functions of the nodal voltage as follows:

$$p = p_0 \left(a v^2 + b v + c \right) \tag{12.22}$$

$$q = q_0 \left(a_q v^2 + b_q v + c_p \right) \tag{12.23}$$

where p_0, q_0, and v represent the nominal active/reactive power and the nodal voltage in per unit. Each coefficient in the quadratic form has a physical meaning according to the type of load, thus c_p and c_q represent constant-power loads whereas coefficients of the linear part (b_p and b_q) represent constant-current loads and, coefficient associated to the quadratic terms (a_p and a_q) represent constant-impedance loads. This model is known as ZIP model since each term represent the percentage of constant impedance loads (Z), constant current loads, (I) and constant power loads (P). Therefore, all coefficients in the model are positive and the following constraints hold:

$$a_p + b_p + c_p = 1 \tag{12.24}$$

$$a_q + b_q + c_q = 1 \tag{12.25}$$

Every load exhibits a mixture of such voltage-dependent behavior. However, constant impedance is most commonly found in residential loads, constant current in commercial loads and, constant power in industrial loads. The use of Advanced Metering Infrastructure (AMI) for real-time monitoring and control of a distribution system allows estimating the model ZIP of the load accurately by a simple optimization problem.

Let us consider a database with measures of nodal voltage and active/reactive power. Our objective is to adjust these measures to a ZIP model taking into account the previously mentioned physical constraints. Therefore, the following least-squares estimation model is proposed for the case of the active power[3]:

---

3 The model for the reactive power has the same structure.

$$\min \frac{1}{2} \sum_{k=0}^{n-1} \left(a_p v_k^2 + b_p v_k + c_p - r_p p_k \right)^2$$

$$a_p + b_p + c_p = 1 \qquad\qquad (12.26)$$

$$a_p, b_p, c_p, r_p \geq 0$$

where p_k and v_k are measures of power and nodal voltage with $\dim(p) = \dim(v) = n$ and $r_p = 1/p_0$. The objective is to minimize the sum of the squares of the error between the data and the model, subject to (12.24). Notice that Model (12.26) is convex and can be easily solved using CvxPy. This model can be executed for a database grouped by hours, in order to obtain values of each hour of the day, but can be also executed to evaluate the average behavior of the load.

Example 12.7. Let us see how Model (12.26) works in practice; first, we generate a set of $n = 800$ synthetic measurements of voltage and active/reactive power. Voltages are randomly generated through a normal distribution such that most of the data is between 0.95 and 1.05. To do this, we define a mean of 1 and a standard deviation of $0.05/3$ (thus, the 98% of the data falls into this interval). Active and reactive power are calculated by a predefined ZIP model with additional noise, also generated by a normal distribution as follows:

```
import numpy as np
n = 800
v = np.random.normal(1,0.05/3,n)
p = 1.2*(0.3*v**2+0.2*v+0.5) + np.random.normal(0,0.004,n)
q = 0.2*(0.6*v**2+0.2*v+0.2) + np.random.normal(0,0.001,n)
```

Figure 12.5 shows the results for one execution of this code.

A script for the estimation of the ZIP model for the active power is straightforward for this case:

```
ap = cvx.Variable(nonneg=True)
bp = cvx.Variable(nonneg=True)
cp = cvx.Variable(nonneg=True)
rp = cvx.Variable(nonneg=True)
fo = cvx.Minimize(1/2*cvx.sum((ap*v**2+bp*v+cp-rp*p)**2))
re = [ap+bp+cp == 1]
Model = cvx.Problem(fo,re)
Model.solve()
```

The student is invited to execute the script and evaluate the accuracy of the estimation, taking into account this synthetic data set is perhaps more distorted than actual measurements. More general models are possible, including the

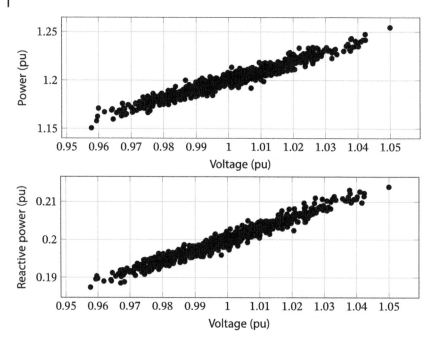

Figure 12.5 Synthetic data of voltage and power for load estimation.

effect of the frequency. The structure of these models is also the least square model and can be solved quickly, as presented here.

Example 12.8. The least square model for the load estimation can be solved directly by relaxing inequality constraints. In that case, the Lagragian is given by (12.27):

$$\mathcal{L} = \frac{1}{2} \sum_{k=0}^{n-1} \left(a_p v_k^2 + b_p v_k + c_p - r_p p_k\right)^2 + \lambda_p(a_p + b_p + c_p - 1) \quad (12.27)$$

The optimal conditions can be easily obtained by taking the derivative of $\mathcal{L}$ as function of a_p, b_p, c_p, r_p, and λ_p, resulting in the following linear system:

$$\begin{pmatrix} \sigma_{v4} & \sigma_{v3} & \sigma_{v2} & -\sigma_{pv2} & 1 \\ \sigma_{v3} & \sigma_{v2} & \sigma_{v1} & -\sigma_{pv1} & 1 \\ \sigma_{v2} & \sigma_{v1} & n & -\sigma_{p1} & 1 \\ \sigma_{pv2} & \sigma_{pv1} & \sigma_{p1} & -\sigma_{p2} & 0 \\ 1 & 1 & 1 & 0 & 0 \end{pmatrix} \begin{pmatrix} a_p \\ b_p \\ c_p \\ r_p \\ \lambda_p \end{pmatrix} = \begin{pmatrix} 0 \\ 0 \\ 0 \\ 0 \\ 1 \end{pmatrix} \quad (12.28)$$

with

$$\sigma_{v4} = \sum_{k=0}^{n-1} v_k^4 \qquad \sigma_{v3} = \sum_{k=0}^{n-1} v_k^3$$

$$\sigma_{v2} = \sum_{k=0}^{n-1} v_k^2 \qquad \sigma_{v1} = \sum_{k=0}^{n-1} v_k$$

$$\sigma_{pv2} = \sum_{k=0}^{n-1} p_k v_k^2 \qquad \sigma_{pv1} = \sum_{k=0}^{n-1} p_k v_k \qquad (12.29)$$

$$\sigma_{p1} = \sum_{k=0}^{n-1} p_k \qquad \sigma_{p2} = \sum_{k=0}^{n-1} p_k^2$$

By solving this five linear system, we may obtain the values of a_p, b_p, c_p, and r_p; however, in some pathological cases, the values may be negatively violating the inequality constraint.

12.6 Further readings

One of the first formulations of the state estimation problem can be found in [106]; see also [103] and the references therein for a complete review of the classic formulation. A modern approach considering PMUs can be found in [107]. In general, the problem is non-convex. However, there are convex approximations including semidefinite programming [108], just as in the case of the power flow equations [109]. The problem may be complemented with other algorithms that check connectivity and observability of the grid as well as the presence of insufficient data as presented in [110].

Although methods for grid identification have been known for a long time, their application is recently enhanced by the development of PMUs. Modern approaches to the problem are based on statistical algorithms such as the least absolute shrinkage and selection operator (LASSO) proposed in [104], where the problem is referred to as the inverse power flow.

Topology identification is an active research area with different models according to the type of available measurement. The model presented here was proposed by Farajollahi et al. in [105]. This model is attractive for practical application since it requires very few low-cost measurements. In addition, it is notably robust to the variations of the pseudo-measurements.

Parameter identification for aggregate load modeling has been studied under different approaches. For example, in [111] a hybrid learning algorithm was proposed. This algorithm combines heuristics with a non-linear Levenberg–Marquardt method and allows for the representation of static loads such as induction motors and residential, commercial, and industrial loads, as presented in this chapter. The advantages of heuristic algorithms compared to more simple least square methods are questionable in time calculation. However, the current development of machine learning and artificial intelligence is based on these types of approaches, and the research continues.

It is vital to note that although the mathematical models presented in this chapter were solved using CvxPy, they could be solved using the gradient method or by direct calculation of the quadratic problem. These methods may be more efficient for online estimation where results are required in real-time. However, the approach presented here is enough for off-line applications even with large data sets.

12.7 Exercises

1. Solve the dc state estimation problem presented in Example 12.1 including PMUs measurements of the angles of the system, that is, including a new set of measurements $z_\theta = (\theta_0, \theta_1, \theta_2)^T + (e_{\theta_1}, e_{\theta_2}, e_{\theta_2})^T$ with $\sigma_\theta = 1 \times 10^{-6}$. Compare the results with the classic problem without PMUs.
2. Solve the ac state estimation problem presented in Example 12.2, including nodal and branch power flows measurements. Formulate a second-order approximation to make the problem convex.
3. Repeat the previous problem using a semidefinite approximation.
4. Solve the non-convex formulation of the state estimation problem, considering nodal and branch power flows measurements, using Newton's method. Compare results with the previous problems.
5. Solve the topology identification problem on the IEEE 33-bus test system presented in Table 11.2, Chapter 11.
6. Prove that the matrix norm defined as (12.18) has the following property:

$$\|A \cdot B\| \le \|A\| \cdot \|B\| \tag{12.30}$$

7. The gradient method may be also used in matrix functions. Consider the following optimization problem:

$$\min f(X) = \frac{1}{2} \|AX - B\|_F^2 \tag{12.31}$$

where A, B, X are square matrices and $\|\cdot\|_F$ is the Frobenius norm, defined as (12.32).

$$\|X\| = \sqrt{\text{tr}(XX^T)} \tag{12.32}$$

The derivative of this norm is given by (12.33),

$$\frac{\partial \|X\|_F^2}{\partial X} = 2X \tag{12.33}$$

Notice this is the derivative of a matrix since the gradient method must be formulated in terms of matrices and not vectors. Solve the optimization problem (12.31) for two randomly generated matrices $A, B \in \mathbb{R}^{3 \times 3}$.

8. Prove the properties of the Y_{bus} given in (12.21).

9. Repeat examples presented in Section 12.4 but this time use the Frobenius norm. Compare results and computation time.

10. Solve the problem in Example 12.7 by direct derivation of (12.27). Compare the solution with the solution given by CvxPy at different randomly generated instances of the problem.

13

Demand-side management

Learning outcomes

By the end of this chapter, the student will be able to:

- To formulate basic models for demand-side management, including energy storage devices.
- To discuss the effect of electric vehicles.
- To solve problems related to phase balancing and load shifting.

13.1 Shifting loads

Final-users usually pay for consumed energy regardless of the shape of the load curve, that is to say, they pay for kWh and not for kW. However, power is becoming more critical in modern electric markets where the price of the energy is variable along the day. In these cases, it is advisable to adjust the load curve in order to avoid a load peak at high-price hours. Consider, for example, the duty cycle of a washing machine on a residential user; it is preferable to have a peak of demand in hours when the price of energy is low, early in the morning, instead of having it when the price is high in the evening. The same applies to many industrial loads that may be moved in time to reduce the total energy cost.

An optimization model for shifting loads modifies the starting time but maintains the shape and the total energy as depicted in Figure 13.1. This shifting action may be represented as a cyclic permutation.

Let us consider a shifting of one hour. The load curve is represented as a vector $s \in \mathbb{R}^{24}$ for each hour t; the shifted load represented by another vector

Mathematical Programming for Power Systems Operation: From Theory to Applications in Python. First Edition. Alejandro Garcés.
© 2022 by The Institute of Electrical and Electronics Engineers, Inc. Published 2022 by John Wiley & Sons, Inc.

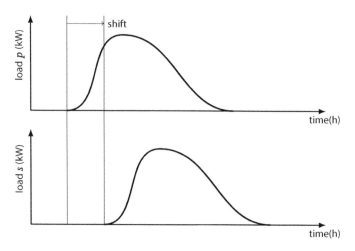

Figure 13.1 Example of a shifting load. In both cases the shape and the total energy are the same, but the peak as moved.

$p \in \mathbb{R}^{24}$, and permutation matrix M which is given by Equation (13.1):

$$M = \begin{pmatrix} 0 & 1 & 0 & 0 & ... & 0 & 0 \\ 0 & 0 & 1 & 0 & ... & 0 & 0 \\ 0 & 0 & 0 & 1 & ... & 0 & 0 \\ 0 & 0 & 0 & 0 & ... & 0 & 0 \\ \vdots & \vdots & \vdots & \vdots & & \vdots & \vdots \\ 0 & 0 & 0 & 0 & ... & 0 & 1 \\ 1 & 0 & 0 & 0 & ... & 0 & 0 \end{pmatrix} \tag{13.1}$$

This matrix shifts the load in one position, that is it returns the same load curve but delayed in one hour; thus, the shifted load is a vector given by the expression $p = Ms$, which returns $p_2 = s_1$, $p_3 = s_2$, $p_4 = s_3$ and so on, until $p_{24} = s_{23}$ and finally, $p_1 = s_{24}$. If we desire to shift the load in two positions, then we must use twice the same permutation, namely $p = MMs = M^2s$. In general, a shift permutation of k positions is given by Equation (13.2):

$$p = M^k s \tag{13.2}$$

Therefore, we must choose among 24 different cyclic permutation matrices including the identity permutation[1]. With this simple idea, the model for shifting load optimization of $\mathcal{D}$ loads is the following:

---

1 Notice that $M^{24} = $ Identity

$$\min c^\top p_{\text{total}}$$

$$p_{\text{total}} = \sum_{i \in \mathcal{D}} p_i$$

$$p_i = \left(\sum_k x_{ik} M^k \right) s_i, \quad \forall i \in \mathcal{D} \tag{13.3}$$

$$\sum_k x_{ik} = 1, \quad \forall i \in \mathcal{D}$$

$$x_{ik} \in \{0, 1\}$$

where c is the vector of energy cost for 24 h operation; x_{ik} is a binary variable which is 1 if the load i is shifted to k positions; and M is the one-shift permutation given by Equation (13.1). The objective function seeks to minimize total costs, subject to the energy balance. Additional constraint may be included, such as maximum peak load $p_{\text{total}} \leq p_{\max}$ and hours of banned operation (e.g, we might avoid operation between 0:00 and 7:00 AM). Notice that Equation (13.3) is a mixed-integer programming problem, since M^k is constant.

Example 13.1. Let us make a small experiment to see the properties of the cyclic permutation matrices. In the script below, we create the matrix M and a permutation M^k with k as input. The result is the expected permutation according to k.

```python
import numpy as np
M = np.zeros((24,24))
M[23,0]  = 1
for k in range(23):
    M[k,k+1] = 1
t = int(input('Enter the time shifting:'))
p = np.linspace(1,24,24)
R = np.linalg.matrix_power(M,t)
m = p @ R
for k in range(24):
    print(p[k],'->',m[k])
```

The student is invited to use different integer values of k, even values greater than 24.

Example 13.2. An industry has three specific loads that correspond to a different industrial process. These loads can be shifted in order to minimize total costs. The capability of the transformer is $p_{\max} = 20$ and no process can be performed between 0:00 and 6:00 AM. Table 13.1 shows the data for the current operation.

Table 13.1 Three industrial process and price of the energy in each time.

Hour	Price	Load 1	Load 2	Load 3
0	460	0	0	0
1	450	0	0	0
2	450	0	0	0
3	450	0	0	0
4	470	0	0	0
5	470	0	0	0
6	500	0	0	0
7	530	0	0	2
8	580	5	0	3
9	600	10	3	4
10	620	10	4	10
11	600	8	10	12
12	600	5	12	7
13	600	3	9	5
14	580	0	8	4
15	570	0	5	0
16	560	0	0	0
17	565	0	0	0
18	550	0	0	0
19	610	0	0	0
20	650	0	0	0
21	650	0	0	0
22	610	0	0	0
23	500	0	0	0

The following code shows the implementation of Model Equation (13.3) for these three loads. First, we load the data, which was stored in a CSV file named `Table.csv`.

```
import numpy as np
import matplotlib.pyplot as plt
from pandas import read_csv
import cvxpy as cvx
```

```
data = read_csv('Table.csv')
c = data['Price'].to_numpy(dtype= float)
s1 = data['Load 1'].to_numpy(dtype= float)
s2 = data['Load 2'].to_numpy(dtype= float)
s3 = data['Load 3'].to_numpy(dtype= float)
```

Now, we define the matrix M:

```
M = np.zeros((24,24))
M[23,0]  = 1
for k in range(23):
    M[k,k+1] = 1
```

Finally, we built the optimization model:

```
x1 = cvx.Variable(24,boolean=True)
x2 = cvx.Variable(24,boolean=True)
x3 = cvx.Variable(24,boolean=True)
pt = cvx.Variable(24)
p1 = cvx.Variable(24)
p2 = cvx.Variable(24)
p3 = cvx.Variable(24)
u = np.ones(24)
pmax = 20
Eq1 = 0
Eq2 = 0
Eq3 = 0
for k in range(24):
    Eq1 = Eq1 + x1[k]*np.linalg.matrix_power(M,k)@s1
    Eq2 = Eq2 + x2[k]*np.linalg.matrix_power(M,k)@s2
    Eq3 = Eq3 + x3[k]*np.linalg.matrix_power(M,k)@s3
res = [pt == p1+p2+p3,
        pt <= pmax,
        p1 == Eq1,
        p2 == Eq2,
        p3 == Eq3,
        cvx.sum(x1) == 1,
        cvx.sum(x2) == 1,
        cvx.sum(x3) == 1,
        pt[0:6] == 0]
fo = cvx.Minimize(c.T@pt)
SL = cvx.Problem(fo,res)
SL.solve()
```

Most of the code is self-explanatory. Notice there are only 24 binary variables for each load in this Model. The rest of the variables are continuous.

Results are shown in Figure 13.2. Initial loads are shown at the top, while shifted loads are shown at the bottom. Notice the peak of the load was 30 kW in the first case, while the new peak is only 20 in the shifted loads. The shifting

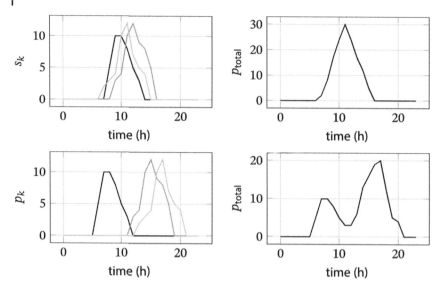

Figure 13.2 Results for the shifting load problem. Original loads (top), shifted loads (bottom).

load optimization reduces by 5% the total cost. A great reduction taking into account we only required to move the loads in time.

13.2 Phase balancing

Power distribution networks are usually unbalanced due to the presence of single-phase loads. This phenomenon is undesired since unbalancing increases neutral currents and, consequently, power losses. In addition, unbalanced currents may reduce motors lifetime since extra heat, due to zero-sequence currents, increases operating temperature of windings and might break down insulation, entailing motor failure. Mechanical vibrations due to unbalanced voltages/currents in industrial loads may also reduce the lifetime of motors. Therefore, we require an optimization model to determine the phase in which loads are connected to reduce zero sequence currents [112].

Let us consider a system with n single-phase loads represented as d_j with $j \in \{0, 1, \dots, n-1\}$, which are connected to a three-phase power distribution network; the problem is basically an assignment problem[2] that consists

---

2 See [113] for more details about the general assignment problem.

Figure 13.3 Phase balancing as an assignment problem.

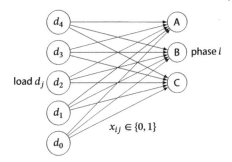

in assigning a phase to each load as shown schematically in Figure 13.3. Therefore, a binary variable x_{ij} is defined such that $x_{ij} = 1$ if the load d_j is connected to the phase $i \in \{A, B, C\}$. The total power in each phase is given the matrix equation Equation (13.4).

$$\begin{pmatrix} p_A \\ p_B \\ p_C \end{pmatrix} = \begin{pmatrix} x_{A0} & x_{A1} & \cdots & x_{An-1} \\ x_{B0} & x_{B1} & \cdots & x_{Bn-1} \\ x_{C0} & x_{C1} & \cdots & x_{Cn-1} \end{pmatrix} \begin{pmatrix} d_0 \\ d_1 \\ \vdots \\ d_{n-1} \end{pmatrix} \tag{13.4}$$

Each load must be connected to a unique phase, therefore, the following constraint must be satisfied:

$$\sum_{i \in \{A,B,C\}} x_{ij} = 1 \quad \forall j \in \{0, 1, \dots, n-1\} \tag{13.5}$$

An unbalance index ψ is defined as the deviation of the power in each phase with respect to the mean:

$$\psi = \frac{1}{3} (\|p_A - p_{\text{mean}}\| + \|p_B - p_{\text{mean}}\| + \|p_C - p_{\text{mean}}\|) \tag{13.6}$$

where p_A, p_B, p_C are the total power in each phase and, p_{mean} is the average power per phase given by Equation (13.7),

$$p_{\text{mean}} = \frac{p_A + p_B + p_C}{3} \tag{13.7}$$

under ideal balanced conditions, p_A, p_B, and p_C would be equal and ψ would be zero; however, this condition is not always possible and hence, our best alternative is to find the configuration that minimizes ψ. This configuration may not be unique but the value of ψ does since Equation (13.6) is a strongly convex function. The entire model is presented below:

$$\min \psi = \frac{1}{3} \left(\|p_A - p_{\text{mean}}\| + \|p_B - p_{\text{mean}}\| + \|p_C - p_{\text{mean}}\| \right)$$

$$p_{\text{mean}} = \frac{p_A + p_B + p_C}{3}$$

$$p = xd \qquad\qquad (13.8)$$

$$\sum_{i \in \{A,B,C\}} x_{ij} = 1 \; \forall j$$

$$x_{ij} \in \{0, 1\}$$

where $x, p_A, p_B, p_C,$ and p_{mean} are decision variables and d is the vector of single-phase loads. The model can be easily transformed into a mixed-integer programming problem that can be solved using CvxPy as presented in the following example:

Example 13.3. Let us balance a system with nine single-phase loads given by a vector $d = (5, 8, 7, 3, 5, 5, 3, 4, 4)^\top$ kW. The model consists in Equations (13.4) to (13.7) with x a Boolean matrix:

```
import cvxpy as cvx
import numpy as np
d = np.array([5,8,7,3,5,5,3,4,4]) # single-phase loads
n = len(d)
A, B, C = 0, 1, 2
x = cvx.Variable((3,n), boolean = True)
p = cvx.Variable(3)
pmean = cvx.Variable()
psi = cvx.Minimize(1/3*(cvx.abs(pmean-p[A])+
                    cvx.abs(pmean-p[B])+
                    cvx.abs(pmean-p[C])))
re  = [pmean == sum(p)/3]
re += [p == x@d]
re += [x[A,j] + x[B,j] + x[C,j] == 1 for j in range(n)]
PhaseBalance = cvx.Problem(psi,re)
PhaseBalance.solve()
print(np.round(np.abs(x.value)))
print(p.value,pmean.value)
```

The optimal value is $\psi = 0.444$ with a total power of 15 kW in two of the phases and 14 kW in the remaining phase. This solution is, of course, not unique since there are different manners to obtain the same unbalance index. The student is invited to test different solvers for the Model and create instances with more loads.

The previous Model is valid for single-phase loads. In the case of three-phase loads, the assignment model must take into account that each phase of the load

Table 13.2 Feasible permutations for the phase-balancing problem.

Matrix	Value	Permutation	Determinant
M_0	$\begin{pmatrix} 1 & 0 & 0 \\ 0 & 1 & 0 \\ 0 & 0 & 1 \end{pmatrix}$	ABC	+1
M_1	$\begin{pmatrix} 0 & 1 & 0 \\ 0 & 0 & 1 \\ 1 & 0 & 0 \end{pmatrix}$	BCA	+1
M_2	$\begin{pmatrix} 0 & 0 & 1 \\ 1 & 0 & 0 \\ 0 & 1 & 0 \end{pmatrix}$	CAB	+1
M_3	$\begin{pmatrix} 1 & 0 & 0 \\ 0 & 0 & 1 \\ 0 & 1 & 0 \end{pmatrix}$	ACB	−1
M_4	$\begin{pmatrix} 0 & 1 & 0 \\ 1 & 0 & 0 \\ 0 & 0 & 1 \end{pmatrix}$	BAC	−1
M_5	$\begin{pmatrix} 0 & 0 & 1 \\ 0 & 1 & 0 \\ 1 & 0 & 0 \end{pmatrix}$	CBA	−1

must be unequivocally connected to each phase of the system as was presented in Figure 1.6, Chapter 1. The problem is now a permutation with six possible configurations given in Table 13.2, where each permutation is described by a 3×3 matrix. The first three permutations maintain the sequence of the original loads while the three last permutations reverse the sequence. The determinant of the matrix indicates this property[3]. In general, the problem may be stated in terms of these possible permutations[4].

A set $\mathcal{D}$ is defined to represent the three phase loads that require to be balanced. Each load is vector $d_k \in \mathbb{R}^3$ with $k \in \mathcal{D}$. For the sake of simplicity,

---

3 $\det(M_i) = 1$ if the permutation maintains the sequence. A change in the sequence of the load may have practical effects. For example, a reverse in the sequence of a motor causes an opposite rotation.

4 Other representations are possible. However, this representation reduces the number of binary variables. See Exercise 6 at the end of this chapter.

we assume only real power, but the model can be extended to consider active and reactive power. A new vector of loads p_k is defined with the required permutation as given below:

$$p_k = (x_{0k}M_0 + x_{1k}M_1 + x_{2k}M_2 + x_{3k}M_3 + x_{4k}M_4 + x_{5k}M_5)\,d_k \qquad (13.9)$$

where x_{ik} is a binary variable which is 1 as the permutation i is activated in the load j. This decision must be univocal, so the following constrain must be added:

$$x_{0k} + x_{1k} + x_{2k} + x_{3k} + x_{4k} + x_{5k} = 1 \qquad (13.10)$$

Finally, the power per phase is represented by a vector $s \in \mathbb{R}^3$ and the average power is represented by s_{mean}. The mathematical model is given below:

$$\min \; \psi = \frac{1}{3}\left(\sum_{k \in \Omega} |s_{\text{mean}} - s_k|\right)$$

$$p_k = \sum_{i=0}^{5} x_{ik}M_i d_k, \quad \forall k \in \mathcal{D}$$

$$\sum_{i=0}^{5} x_{ik} = 1, \quad \forall k \in \mathcal{D}$$

$$s_{\text{mean}} = \frac{1}{3}\sum_{k \in \Omega} s_k \qquad (13.11)$$

$$s_k = \mathbf{1}^T p$$

$$x_{ik} \in \{0, 1\}$$

where $\Omega = \{A, B, C\}$ represents the phases of the system. Notice this is a integer linear programming model that can be easily coded in CvxPy. This is only a toy-model, since the phase balance problem may be integrated with the power flow equations to minimize power loss in power distribution systems. Interested reader is referred to [114].

Example 13.4. Let us make an instance of Model Equation (13.11) for a system with four loads $d_0 = (10, 12, 11)^T$, $d_1 = (12, 15, 11)^T$, $d_2 = (11, 12, 15)^T$, $d_3 = (13, 10, 12)^T$. First, we define a list that contains the permutation matrices given in Table 13.2:

```
import numpy as np
import cvxpy as cvx
M = []
M += [np.array([[1,0,0],[0,1,0],[0,0,1]])]
M += [np.array([[0,1,0],[0,0,1],[1,0,0]])]
M += [np.array([[0,0,1],[1,0,0],[0,1,0]])]
```

```
M += [np.array([[1,0,0],[0,0,1],[0,1,0]])]
M += [np.array([[0,1,0],[1,0,0],[0,0,1]])]
M += [np.array([[0,0,1],[0,1,0],[1,0,0]])]
```

Now, we define the optimization model whose implementation is straightforward:

```
d = np.array([[10,12,11,13],
              [12,15,12,10],
              [11,11,15,12]])

num_loads = len(d.T)
x = cvx.Variable((6,num_loads),boolean=True)
p = cvx.Variable((3,num_loads))
s_mean = cvx.Variable()
s = cvx.Variable(3)
res = []
for k in range(num_loads):
    res += [p[:,k] == (x[0,k]*M[0] +
                       x[1,k]*M[1] +
                       x[2,k]*M[2] +
                       x[3,k]*M[3] +
                       x[4,k]*M[4] +
                       x[5,k]*M[5])@d[:,k]]
    res += [cvx.sum(x[:,k]) == 1]

res += [s_mean==sum(s)/3]
res += [s == sum(p.T)]

obj = cvx.Minimize(1/3*(cvx.abs(s_mean-s[0])+
                        cvx.abs(s_mean-s[1])+
                        cvx.abs(s_mean-s[2])))
PhaseBalancing = cvx.Problem(obj,res)
PhaseBalancing.solve()
```

Finally, we print the result with the following code:

```
print(s.value)
u = np.array([0,1,2])
pha = 'ABC'
for k in range(num_loads):
    for i in range(6):
        if np.round(x[i,k].value) == 1:
            Mk = M[i]
            w = Mk@u
            print(pha[w[0]],pha[w[1]],pha[w[2]])
```

This result could have been obtained by hand since the problem is quite simple. However, the model is general for any number of loads. The student is invited to generate more instances of the model with a high number of loads.

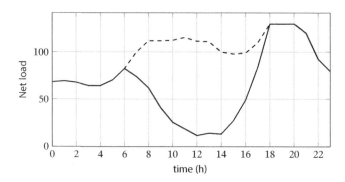

Figure 13.4 Load curve in a typical day: (- -) normal load without solar generation, (—) load curve including the effect of solar generation (duck curve).

13.3 Energy storage management

High levels of wind and solar generation lead to a drastic change in the load curve. For example, a peak of solar radiation is expected in the middle of the day, creating a drastic reduction of the demand seen by the power distribution system. Likewise, the peak of the load is usually expected in the late evening when the capacity for solar generation is reduced. This produces a steep slope of the load as shown in Figure 13.4, a phenomenon known as the *duck curve*[5]. This creates problems for the power system operator, especially in the late evening when demand begins to rise since this steep slope affects the unit commitment and the economic dispatch.

One way to mitigate problems related to the duck curve is using energy storage devices such as batteries, flywheels, compressed air, or superconducting magnetic energy storage. The cycles of charge/discharge of these devices should be optimized to reduce the adverse effects of the duck curve. They can also be optimized to reduce power loss or to increase profits for selling energy to the grid. All cases result in a convex optimization problem that can be easily solved in Python.

Here, we propose a toy-model for the ideal grid-connected microgrid with solar system and energy storage depicted in Figure 13.5. This model, although particular, may be easily modified to include more general phenomena and components. Our main objective is to minimize cost, although other objectives such as peak shaving or loss minimization are also viable.

---

5 This term was coined by the California Independent System Operator but now used in the power system literature.

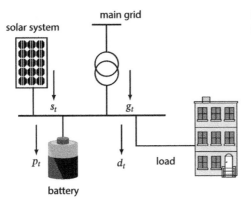

Figure 13.5 Example of a grid-connected microgrid.

Let us define s_t as the power generated by the solar system and g_t as the power injected by the main grid; p_t is the power required by the storage system (in this case a battery) and d_t is the power required by the load. Therefore, the balance of power in the common bus-bar is given by Equation (13.12):

$$s_t + g_t = p_t + d_t \tag{13.12}$$

where the sub-index t indicates the time in hours for a operation in 24 h. Power is measured in kW, time in h and energy in kWh; thus, the energy in the next hour is given by Equation (13.13):

$$e_{t+1} = e_t + p_t \tag{13.13}$$

By convention we assume $p_t > 0$ for charging mode and $p_t < 0$ for discharging mode. For the sake of simplicity we assume an ideal process with 100% efficiency. Both the price of the energy c_t, the power generated by the solar system s_t and, the load d_t are known a priory via an accurate forecasting, so the optimization model is a deterministic linear programming problem:

$$\min \sum_t c_t g_t$$
$$g_t + s_t = p_t + d_t$$
$$e_{t+1} = e_t + p_t$$
$$0 \leq e_t \leq e^{\max} \tag{13.14}$$
$$|p_t| \leq p^{\max}$$
$$e_0 = 0$$

As mentioned before, the objective is to minimize total costs, and the main result is the schedule for charge and discharge of the battery. The following example shows the use of the Model in practice.

Example 13.5. Table 13.3 shows a forecasting for solar generation, load and price for a microgrid as the one shown in Figure 13.5. This table is stored in a cvs file named `Table.csv`. We are going to use data frames in Pandas as a tool for analysis and data manipulation.

Model Equation (13.14) is implemented in Python with $e^{max} = 100\,\text{kWh}$ and $p^{max} = 30\,\text{kW}$. The battery start discharged at $t = 0$. The corresponding code is presented below:

```python
import numpy as np
import matplotlib.pyplot as plt
import cvxpy as cvx
from pandas import read_csv
data = read_csv('Table.csv')
p_stor = cvx.Variable(24)   # p>0 charging
p_grid = cvx.Variable(24)   # p<0 discharging
e_stor = cvx.Variable(25)
pmax = 30
emax = 100
f = 0
res = [p_stor>=-pmax,
       p_stor<= pmax,
       e_stor>=0,
       e_stor<=emax,
       e_stor[0]==0]
for t in range(24):
    f += data['Price'][t]*p_grid[t]
    res += [e_stor[t+1] == e_stor[t] + p_stor[t]]
    res += [p_grid[t]+data['Solar'][t]==p_stor[t]+data['Load'][t]]
BM = cvx.Problem(cvx.Minimize(f),res)
BM.solve()
```

Results for 24 h operation are plotted with the following code:

```python
plt.subplot(3,1,1)
data['Solar'].plot()
data['Load'].plot()
plt.plot(p_stor.value)
plt.plot(p_grid.value)
plt.legend(['Solar','Load','Battery','Grid'])
plt.grid()
plt.ylabel('Power (kW)')
plt.subplot(3,1,2)
plt.plot(e_stor.value)
plt.grid()
plt.ylabel('Energy (kWh)')
```

```
plt.subplot(3,1,3)
data['Price'].plot()
plt.ylabel('Spot price ($/kW)')
plt.grid()
plt.show()
```

As common along with the book, the reader is invited to do experiments with this code. For example, compare results for larger values of p^{max} and e^{max}. Compare the results with and without energy storage.

13.4 Further readings

One of the first models for demand-side management can be found in [115], while a modern vision of the problem can be found in [116] and in [117]. An interesting model for local flexibility markets can be found in [33]. Besides, a good review of shifting load models is available in [118].

There are different technologies for energy storage, each one with a particular niche market. For example, batteries are used for energy management, while flywheels are used for stability purposes. A complete review about current technologies for energy storage can be found in [59].

The classic approach for phase balancing in power distribution grids is based on mixed-integer programming as in [114]. However, the problem is more complex when it considers the effects of the power distribution network and the non-linear model of loads. In that case, a mixed-integer non-linear programming model is required. This type of model is usually solved by heuristic algorithms such as simulated annealing [119]. A significant challenge of heuristic algorithms in this type of applications is the representation of the solutions; recent studies have demonstrated the advantages of using group-based codification that reduces the space of solutions. Interested readers are invited to review [120], and the references therein, for a complete analysis of this type of codification.

13.5 Exercises

1. Modify the Model presented in Example 13.5 for peak shaving. Compare results with the case of cost minimization.
2. Evaluate Example 13.5 with different values of p^{max} and e^{max} for both cost minimization and peak shaving. What is more important in each case, to have a large energy storage capacity or a large power capacity?.

Table 13.3 Expected generation, demand and price for 24h operation of a microgrid.

Hour	Solar	Load	Price
0	0	68.5	140
1	0	69.5	140
2	0	68	140
3	0	64.5	140
4	0	64.5	140
5	0	70.5	175
6	0	82.5	210
7	26	100	210
8	50	112	210
9	71	112	175
10	87	112.5	170
11	97	115.5	170
12	100	111.5	173
13	97	111	175
14	87	100	180
15	71	98	190
16	50	99	210
17	26	110	315
18	0	130	385
19	0	130	385
20	0	130	350
21	0	120	280
22	0	92.5	245
23	0	79.5	175

3. Modify the code presented in Example 13.1 to show the cyclic permutation if the problem is divide into 15-minute intervals.
4. Modify Example 13.2 for minimization of the peak load.
5. Generate random instances with different sizes of the phase balancing problem presented in Example 13.3. Evaluate the quality of the solution and time calculation according to the size of the problem. Use different solvers.

6. The phase balancing problem for three-phase loads may be represented without using variables x_{ik} of Model Equation (13.11). In this case, we define a matrix variable $M_k \in \mathbb{B}^{3\times3}$ for each load, such that

$$p_k = M_k d_k \tag{13.15}$$

with additional constraints on the entries m_{ij} of each matrix M_k:

$$\sum_i m_{ij} = 1 \tag{13.16}$$

$$\sum_j m_{ij} = 1 \tag{13.17}$$

Formulate the phase balancing problem in these terms and compare the solutions. How many binary variables are required in this formulation compared to Model Equation (13.11)?

7. Solve the phase balancing problem for three-phase loads avoiding permutations that reverse the sequence.

8. Solve the phase balancing problem relaxing the binary variables. Use a randomly generated instance of the problem with more than ten loads.

9. Formulate the phase balancing problem considering active and reactive power. Propose a suitable objective function in this case.

10. Demand-side management is a vast area of research; there are many other problems such as thermal loads and V2G that are closely related. Search on the internet for these problems and formulate their corresponding models.

A

The nodal admittance matrix

We require some concepts from graph theory in order to obtain a systematic representation of the network equations, and solve large optimization problems such as the optimal power flow. A graph is a structure $\mathcal{G} = \{N, B\}$ that groups together a set of nodes $N = \{0, 1, \ldots, n-1\}$ and a set of edges (branches) $B \subseteq N \times N$ that connect these nodes. For example, Figure A.1 depicts a graph with nodes $N = \{0, 1, 2, 3, 4, 5\}$ and branches $B = \{(0, 1), (0, 3), (1, 2), (1, 4), (1, 5), (2, 5), (3, 4), (4, 5)\}$. In this case, the graph is oriented, since the set of branches B defines not only the connectivity between nodes but also a direction of these connections. This is useful to represent an electric network since it allows to define the direction of the current and/or the power flows in the branches.

The connectivity of the graph can be represented by a matrix A known as incidence matrix. This matrix is size $n \times m$ where n is the number of nodes and m the number of branches[1]. Thus, every entry a_{ij} represents the connections of node i with the branch j; in the case of an oriented graph, $a_{ij} = -1$ if the branch j leaves node i and $a_{ij} = 1$ if the branch j arrives to node i; otherwise $a_{ij} = 0$ if the branch j is not connected to node i. For example, the incidence matrix for the graph depicted in Figure A.1 is the following:

Figure A.1 Example of an oriented graph.

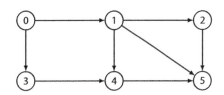

---

[1] Notice that some books define the node–branch incidence matrix which is the transpose of our definition.

Mathematical Programming for Power Systems Operation: From Theory to Applications in Python. First Edition. Alejandro Garcés.

$$A = \begin{pmatrix} -1.0 & -1.0 & 0.0 & 0.0 & 0.0 & 0.0 & 0.0 & 0.0 \\ 1.0 & 0.0 & -1.0 & -1.0 & -1.0 & 0.0 & 0.0 & 0.0 \\ 0.0 & 0.0 & 1.0 & 0.0 & 0.0 & -1.0 & 0.0 & 0.0 \\ 0.0 & 1.0 & 0.0 & 0.0 & 0.0 & 0.0 & -1.0 & 0.0 \\ 0.0 & 0.0 & 0.0 & 1.0 & 0.0 & 0.0 & 1.0 & -1.0 \\ 0.0 & 0.0 & 0.0 & 0.0 & 1.0 & 1.0 & 0.0 & 1.0 \end{pmatrix} \quad (A.1)$$

The incidence matrix is used to define relations between nodal and branch variables in a power grid. For instance, nodal currents can be calculated from branch currents as presented below:

$$I_N = A I_B \quad (A.2)$$

Likewise, branch voltages can be calculated from nodal voltages as follows:

$$V_B = A^T V_N \quad (A.3)$$

Branch currents are in turns related to the branch voltages as given in (A.4) which constitute a matrix representation of the Ohm's law:

$$I_B = Y_B V_B \quad (A.4)$$

where $Y_B = \text{diag}(y_{ij})$ is a diagonal matrix with diagonal entries equal to the admittance of each branch $(ij) \in B$. Let us replace the expressions presented above into (A.4) to obtain a direct relation between nodal currents and nodal voltages:

$$I_N = A Y_B A^T V_N \quad (A.5)$$

From (A.5) we can obtain a direct definition of the nodal admittance matrix Y_{bus}:

$$Y_{bus} = A Y_B A^T \quad (A.6)$$

This is, of course, one of many ways to obtain the Y_{bus} matrix. Another approach consists in defining its entries directly as follows:

$$Y_{bus}(i, i) = \sum_{ij \in \Omega_i} y_{ij} \quad (A.7)$$

$$Y_{bus}(i, j) = -y_{ij} \quad (A.8)$$

where y_{ij} is the admittance of each branch $(ij) \in B$ and Ω_i represents the set of branches that connects the node i. The following example shows how to build the Y_{bus} in practice.

Example A.1. Python allows to calculate easily the incidence matrix of a graph using the module NetworkX. Let us see how the graph of Figure A.1 is represented:

```
import networkx as nx
G = nx.DiGraph()
G.add_nodes_from([0,1,2,3,4,5])
G.add_edges_from([[(0,1),(0,3),(1,2),(1,4),(1,5),(2,5),(3,4),
(4,5)]])
```

Intuitively, we have defined N and B in the last two lines. Thus, we can obtain a representation of the graph using the code given below, which is self-explanatory:

```
import matpLotlib.pyplot as plt
nx.draw(G,with_labels = True)
plt.show()
```

Remember that it is the connections and not the shape of the graph that matter in this figure. Therefore, the draw may look different from Figure A.1, but the graph G is the same.

Example A.2. The incidence matrix can be easily calculated in Python using the modulde NetworkX. Let us continue with the graph depicted in Figure A.1, where the incidence matrix is calculated as given below:

```
A = nx.incidence_matrix(G,oriented = True)
```

Let us suppose the admittance of each branch is $y_{ij} = -10j$, then, the Y_{bus} is obtained as follows:

```
import numpy as np
yB = -10j*np.identity(6)
Ybus = A@yB@A.T
```

Thus, the Y_{bus} is built in only a few code lines.

B

Complex linearization

It is common, in power systems applications, to find equality constraints defined in the complex domain. Representing the equation in the complex domain may turn out to be more straightforward. Compare, for example, the complex representation of the nodal power, presented below:

$$s_k^* = \sum_m y_{km} v_k^* v_m \tag{B.1}$$

with its counterpart separated in real and imaginary part, namely:

$$p_k = \sum_m g_{km} v_k v_m \cos(\theta_{km}) + b_{km} v_k v_m \sin(\theta_{km}) \tag{B.2}$$

$$q_k = \sum_m b_{km} v_k v_m \cos(\theta_{km}) - g_{km} v_k v_m \sin(\theta_{km}) \tag{B.3}$$

Complex equality constraints, such as Equation (B.1), are usually non-convex, so an affine approximation is advisable to convexify the space. A suitable manner to make this approximation is to split the function into real and imaginary parts; then, a truncated Taylor series may be used to obtain an affine function. However, it may be convenient to obtain the approximation directly in the complex domain, as presented in this appendix.

We define the imaginary unit as $j = \sqrt{-1}$; thus, a complex variable is represented as $z = x + jy$, where x and y are the real and imaginary parts, respectively. A function $f : \mathbb{C} \to \mathbb{C}$ is also defined as $f(z) = u + jv$; where $u = \text{real}(f)$ and $v = \text{imag}(f)$. In the context of mathematical optimization, a complex function can be used to represent an equality constraint, as presented below[1]:

$$f(z) = 0 \tag{B.4}$$

---

1 Notice that an inequality constraint, such as $f(z) \leq 0$ may be meaningless in the complex domain, since it is a non-ordered set. See Chapter 2 for a discussion about ordered sets.

Mathematical Programming for Power Systems Operation: From Theory to Applications in Python. First Edition. Alejandro Garcés.
© 2022 by The Institute of Electrical and Electronics Engineers, Inc. Published 2022 by John Wiley & Sons, Inc.

this constraint is equivalent to the following set of constraints in the real domain, namely:

$$u(x,y) = 0$$

$$v(x,y) = 0 \tag{B.5}$$

Obviously, Equation (B.4) is a simpler representation than Equation (B.5). Moreover, Python allows to work directly with complex variables, so it is more convenient to formulate the problem directly in the complex domain. A well-known tool to define a linear approximation is to make use of derivatives. However, a derivative on the complex domain is not as intuitive as a derivative in the real domain.

A derivative, in the complex domain, is defined by the following limit:

$$f'(z) = \lim_{\Delta z \to 0} \frac{f(z + \Delta z) - f(z)}{\Delta z} \tag{B.6}$$

This is the same definition as in the real numbers; however, there are infinitely many directions in which this limit may be taken in the complex plane. This implies an important consideration related to the continuity of the function. In a real function, we require that the limits from the left and the right are equal. Here, we require the same limit from all directions, as depicted in Figure B.1. This fact restricts the analysis to a special set of functions, known as a *holomorphic functions*. Sufficient conditions for differentiating a complex function, are given by the Cauchy–Riemann equations, as presented below:

$$\frac{\partial u}{\partial x} = \frac{\partial v}{\partial y}$$

$$\frac{\partial u}{\partial y} = -\frac{\partial v}{\partial x} \tag{B.7}$$

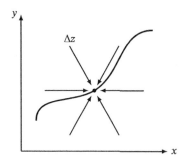

Figure B.1 Possible directions for taking the limit that defines the derivative in the complex plane.

Unfortunately, many expressions in power systems operation problems are not holomorphic. Therefore, we require another mathematical tool that allows us to linearize non-holomorphic functions.

Given a complex function $f = u + jv$ with $u = u(x, y), v = v(x, y)$, we define the Wirtinger's derivative and the conjugate Wirtinger's derivative as follows:

$$\frac{\partial f}{\partial z} = \frac{1}{2}\left(\frac{\partial u}{\partial x} + \frac{\partial v}{\partial y}\right) + \frac{j}{2}\left(\frac{\partial v}{\partial x} - \frac{\partial u}{\partial y}\right) \tag{B.8}$$

$$\frac{\partial f}{\partial z^*} = \frac{1}{2}\left(\frac{\partial u}{\partial x} - \frac{\partial v}{\partial y}\right) + \frac{j}{2}\left(\frac{\partial v}{\partial x} + \frac{\partial u}{\partial y}\right) \tag{B.9}$$

If f is holomorphic, then Wirtinger's derivative is equal to the standard complex derivative. However, in the general case, Wirtinger's derivatives are not the same as the complex derivative.

Both Wirtinger's derivative and conjugate Wirtinger's derivative behave similarly as a partial derivative. So, we can apply common rules for differentiation concerning the sum and the product of functions as follows:

$$\frac{\partial(f + g)}{\partial z} = \frac{\partial f}{\partial z} + \frac{\partial g}{\partial z} \tag{B.10}$$

$$\frac{\partial(f + g)}{\partial z^*} = \frac{\partial f}{\partial z^*} + \frac{\partial g}{\partial z^*} \tag{B.11}$$

$$\frac{\partial(f \cdot g)}{\partial z} = f\frac{\partial g}{\partial z} + g\frac{\partial f}{\partial z} \tag{B.12}$$

$$\frac{\partial(f \cdot g)}{\partial z^*} = f\frac{\partial g}{\partial z} + g\frac{\partial f}{\partial z^*} \tag{B.13}$$

Thus, a linearization is given by the following simple relation that resemblances a Taylor expansion:

$$f(z) \approx f(z_0) + \frac{\partial f}{\partial z}\Delta z + \frac{\partial f}{\partial z^*}\Delta z^* \tag{B.14}$$

Example B.1. Let us formulate a linear approximation for the following function, around the point $z_0 = 0$,

$$f = zz^* + 8z + 5z^* \tag{B.15}$$

First, we calculate the Wirtinger's derivative and the conjugate Wirtinger's derivative, as presented below:

$$\frac{\partial f}{\partial z} = z^* + 8 \tag{B.16}$$

$$\frac{\partial f}{\partial z^*} = z + 5 \tag{B.17}$$

Now, we evaluate these derivatives in the point z_0, obtaining the following affine function:

$$f(z) \approx 0 + 8\Delta z + 5\Delta z^* \tag{B.18}$$

Example B.2. Let us formulate a linear approximation of the function defined in the previous example, but now, we split the function in real and imaginary parts. First, the function is defined as presented below:

$$u(x, y) = x^2 + y^2 + 8x + 5x$$
$$v(x, y) = 8y - 5y \tag{B.19}$$

Now, we define the derivatives in each real variable, namely:

$$\frac{\partial u}{\partial x} = 2x + 13$$

$$\frac{\partial u}{\partial y} = 2y$$

$$\frac{\partial v}{\partial x} = 0 \tag{B.20}$$

$$\frac{\partial v}{\partial y} = 3$$

These derivatives form a jacobian matrix,

$$J = \begin{pmatrix} 2x + 13 & 0 \\ 2y & 3 \end{pmatrix} \tag{B.21}$$

The linear approximation is obtained by a Taylor expansion around $x = 0$ and $y = 0$, that is:

$$\begin{pmatrix} u \\ v \end{pmatrix} = \begin{pmatrix} 0 \\ 0 \end{pmatrix} + \begin{pmatrix} 13 & 0 \\ 0 & 3 \end{pmatrix} \begin{pmatrix} \Delta x \\ \Delta y \end{pmatrix} \tag{B.22}$$

Then, the linear approximation is the following:

$$u = 13\Delta x$$
$$v = 3\Delta y \tag{B.23}$$

Example B.3. Equation (B.18) is equivalent to Equation (B.23); this equivalence is easily demonstrated by the following calculation:

$$8\Delta z + 5\Delta z^* = 8(\Delta x + j\Delta y) + 5(\Delta x - j\Delta y) \tag{B.24}$$

$$= (8\Delta x + 5\Delta x) + j(8\Delta y - 5\Delta y) \tag{B.25}$$

$$= 13\Delta x + 3\Delta y j \tag{B.26}$$

🪱

Example B.4. Power flow equations in a power distribution grid are given by Equation (B.27):

$$s_k^* = \sum_m y_{km} v_k^* v_m \tag{B.27}$$

This equation is non-linear and non-holomorphic. However, we can obtain a linear approximation around $v = 1 + 0j$ as follows:

$$s_k^* = \sum_m y_{km}((1 + 0j)^*(1 + 0j) + (1 + 0j)\Delta v_k^* + (1 + 0j)^*\Delta v_m) \tag{B.28}$$

$$= \sum_m y_{km}(1 + (v_k^* - 1) + (v_m - 1)) \tag{B.29}$$

$$= \sum_m y_{km}(v_k^* + v_m - 1) \tag{B.30}$$

This linearization is convenient for optimization problems such as the optimal power flow, where both s and v are decision variables. Notice that the equation is affine in both variables. 🪱

Example B.5. We can create a different linearization for the power flow equations when s_k is constant. In that case, the nodal current is given by Equation (B.31):

$$i_k = \left(\frac{s_k}{v_k}\right)^* \tag{B.31}$$

This equation is again, non-linear and non-holomorphic. The complex linearization around $v = 1 + 0j$ is the following:

$$i_k = s_k^* \left(\frac{1}{(1 + 0j)^*} - \frac{1}{((1 + 0j)^*)^2}\Delta v_k^*\right) \tag{B.32}$$

$$= s_k^*(1 - (v_k^* - 1)) \tag{B.33}$$

$$= s_k^*(2 - v_k^*) \tag{B.34}$$

This linearization is convenient for problems in which s_k is constant, for example, the primary feeder reconfiguration. 🪱

C

Some Python examples

Python is a powerful programming language for all types of applications, from power systems to game development. There are hundreds of libraries available for free to solve a wide variety of problems. Likewise, there is a vast material available on the internet, such as examples and tutorials. Said tutorials might be more detailed and up-to-date than the examples presented in this appendix, which is only a brief introduction to Python programming.

Python is a high-level and interpreted language, which means the code is executed by an interpreter program, in contrast to compiled languages, such as c++, that return an independent executable. Bypassing the compilation step makes development faster, although the program itself may be a bit slower. The Python interpreter may be downloaded from https://www.python.org/ and works in both Linux and Windows. We do not require anything different from this interpreter to execute the examples presented in this book. The code may be written in a plain document generated in Notepad. However, there are many IDEs (Integrated Development Environments) that simplify the development. We have no preference for any IDE, all of them are good-enough for our purposes. In addition, there are online platforms such as Jupyter and Colab that allow to execute the examples without installing the interpreter. When reading this book, there will probably be many other platforms.

Below, we present a series of basic examples that demonstrate the main features of the language. These examples are pretty simple but enough to understand the logic behind the examples presented in this book.

C.1 Basic Python

Example C.1. Our first example is the traditional hello word program that displays the famous message. Scripting in Python is simple and clean, below, the corresponding code:

Mathematical Programming for Power Systems Operation: From Theory to Applications in Python. First Edition. Alejandro Garcés.
© 2022 by The Institute of Electrical and Electronics Engineers, Inc. Published 2022 by John Wiley & Sons, Inc.

```
print("Hello word")
```

Notice we do not require any additional library or configuration to obtain a
simple output message.

Example C.2. Let us perform simple mathematical operations as follows:

```
x = 5
x = x + 1
print('the value of x is ', x)
```

Other operations such as multiplication and division are straightforward. In
addition, there are commands for floor division and exponentiation, namely:

```
"basic mathematical commands"
x = 10/4    # divide
y = 10//4   # divide and round
z = 10*2    # multiply
w = 10**2   # exponentiation
print(x,y,z,w)
```

Notice that the last case is a square, e.g., $w = 10^2 = 100$. All comments in the
code were done using a hash mark (#).

Example C.3. An array in Python may be stored in different ways; here, we
used manly lists and tuples. A tuple is defined by parenthesis and is immutable,
whereas a list is defined by square brackets and can change in size. See the
difference in the code presented below:

```
X = (10,15,12)   # this is a tuple
Y = [10,15,12]   # this is a list
```

Example C.4. We use tuples when the size of the array is fixed, for example:

```
A = (10,15,12)
```

in this case, A is a vector with three entries. We can access each entry as
`A[0]`, `A[1]`, and `A[2]`. In this case, `A[0]`=10, `A[1]`=15, and `A[2]`=12.
Besides, we can count from the last to the first entry in the following way:
`A[-1]`=12, `A[-2]`=15, and `A[-3]`=10.

Example C.5. We use lists if we require to modify the entries of the array or
its size, for example:

```
A = [10,15,12]
```

We access the entries of A in the same way as a tuple. Moreover, we can increase the size of A, as follows:

```
A += [30]
```

Here, we have added an entry at the end of the list. Therefore, the new list has entries $A[0]=10, A[1]=15, A[2]=12$, and $A[3]=30$.

Example C.6. The operator $*$ is not a conventional multiplication when applied to a list. For instance, the following command returns a vector of size 8 with all entries equal to 5:

```
B = 8*[5]
```

Example C.7. One distinctive characteristic of Python is the indentation, i.e., the spaces at the beginning of a code line. Indentation is used to indicate block code. Let us consider a simple conditional structure:

```
x = 10
y = 20
if x <= y:
    print("x is lower or equal than y")
    print("This line is inside the body of if")
    print("This line is still inside the body of if")
print("This is outside the body of if")
```

Indentation in Python allows a neat code since we do not require any command to begin and end the block code inside the conditional. However, we must be cautious to avoid unnecessary spaces at the beginning of a line code. A simple space may change the results of the algorithm drastically.

Example C.8. Likewise conditionals, a for-loop is quite intuitive in Python. Let us define a simple script that prints the numbers from 0 to 4 and its squares, e.g.,

```
for k in range(5):
    y = k**2
    print('k is ',k,' and k^2 is ',y)
print('This is the end')
```

The first `print` is indented, meaning that this command is executed inside the body of the `for` structure. The second `print` is outside the body of the `for` structure.

Example C.9. A function is defined by the command `def`. The following script shows the definition of the function $f(x) = 1/x^5$:

```
def f(x):
    y = 1/(x**5)
    return y
```

After defining the function, we can evaluate it in any real variable, namely:

```
a = 5.3
b = f(a)
print(b)
```

We can also evaluate the function in a complex number, as follows:

```
a = 5.3 + 2.0j
b = f(a)
print(b)
```

C.2 NumPy

One of the most useful modules in Python is NumPy, which allows operation with multidimensional arrays and matrices similarly to Matlab. The following examples show the use of this module.

Example C.10. The script below, shows a simple definition of a NumPy array:

```
import numpy as np
x = np.array([10,15,12])
```

The first line imports the library and defines an alias (np). The second line defines the array itself; this array behaves as expected in linear algebra. For instance, we can multiply for a scalar as presented below:

```
y = 5*x
```

This operation return a vector $y \in \mathbb{R}^3$ with entries `y[0]=50`, `y[1]=75`, and `y[2]=60`. Note that the result would be very different if x were a list; in that case, the result would be a list of size 15.

Example C.11. A matrix may be easily defined using NumPy. Consider the following 3×3 matrix:

$$A = \begin{pmatrix} 4 & 8 & 7 \\ 3 & 0 & 1 \\ 4 & 2 & 1 \end{pmatrix} \tag{C.1}$$

This array is defined as follows:

```
A = np.array([[4,8,7],[3,0,1],[4,2,1]])
```

Now, we can make different operations related to linear algebra. These operations are available in the `linalg` submodule. Some common functions are presented below:

```
B = np.linalg.inv(A)      # inverse
d = np.linalg.det(A)      # determinant
L,V = np.linalg.eig(A)    # eigenvalues and eigenvectors
```

Example C.12. We can solve a linear system of equations $Ax = b$, were b is a NumPy array of suitable size, for instance:

```
A = np.array([[4,8,7],[3,0,1],[4,2,1]])
b = np.array([12,15,9])
x = np.linalg.solve(A,b)
```

Example C.13. Conventional matrix multiplication is performed by the command @. Let us consider the evaluation of the following quadratic form:

```
x = np.array([1,8,3])
H = np.array([[4,8,7],[3,0,1],[4,2,1]])
f = x.T @ H @ x
```

Example C.14. Conventional mathematical functions are also defined in NumPy, as follows:

```
x = 0.5
a = np.sin(x)
b = np.cos(x)
```

```
c = np.tan(x)
d = np.exp(x)
```

C.3 MatplotLib

Example C.15. MatplotLib is a library that allows to obtain plots in a way as simple as Matlab. The code below, shows the plot of the function $f(x) = \sin(x)/x$ for $-10 \le x \le 10$:

```
import numpy as np
import matplotlib.pyplot as plt
xr = np.linspace(-10,10,100) # vector with 100 points
     from -10 to 10
yr = np.sin(xr)/xr
plt.plot(xr,yr)
```

the command `linspace` create a vector with 100 points, between -10 and 10; next, the function f is invoked and the function is plotted. After that, we can add some labels to the axis, as follows:

```
plt.grid()
plt.xlabel('abscissa')
plt.ylabel('ordinate')
plt.show()
```

C.4 Pandas

Most of the examples presented in this book are toy-models. However, they can be extended to solve large models. In that case, we require a simple and efficient way to read, store, and manipulate data. The module Pandas allows these operations.

Example C.16. The essential component of Pandas is a DataFrame which allows to store and manipulate data. The following code shows the creation of a DataFrame that store the information given in Table C.1:

```
import pandas as pd

mytable = pd.DataFrame()
mytable["Source"] = ["Solar","Wind","Hydro","Geothermal"]
mytable["Installed"] = [12.1, 61.1, 78.4, 3.4]
```

```
mytable["Increasing"] = [10.9, 26.0, 7.3, 0.2]
mytable["Percentage"] = [1.14, 5.76, 7.69,0.32]
mytable.head()
```

This Table may be also stored in a CSV file. In that case, we can open the file by a simple line of code as presented below:

```
mytable = pd.read_csv("MYFILE.csv")
```

where the table is stored in a file named `MYFILE.csv` inside the same folder of the script. The following line returns the source in the second row (i.e., wind):

```
print(mytable["Source"][1])
```

Table C.1 Comparison of the installed power and increase in the United States from 2008 to 2013. Taken from [121]

Source	Installed	Increasing	Percentage
Solar	12.1	10.9	1.14
Wind	61.1	26.0	5.76
Hydro	78.4	7.3	7.39
Geothermal	3.4	0.2	0.32

We can also plot the information given in the DataFrame using MatplotLib, as follows:

```
import matplotlib.pyplot as plt
mytable.plot()
plt.grid()
plt.show()
```

There are many functionalities available in Pandas. This is just a hint about the possibilities of the module. As always, the reader is invited to explore further functions.

Bibliography

[1] Stanfield S, Safdi S, Mihaly S. Optimizing the grid, a regulator's guide to hosting capacity analyses for distributed energy resources. 1st ed. NY: IREC; 2017.

[2] Ding F, Mather B. On Distributed PV Hosting Capacity Estimation, Sensitivity Study, and Improvement. IEEE Transactions on Sustainable Energy. 2017 July;8(3):1010–1020.

[3] Conejo A, Baringo L. Power system operations. springer; 2018.

[4] Terorde M, Wattar H, Schulz D. Phase balancing for aircraft electrical distribution systems. IEEE Transactions on Aerospace and Electronic Systems. 2015 July;51(3):1781–1792.

[5] Weckx S, Driesen J. Load Balancing With EV Chargers and PV Inverters in Unbalanced Distribution Grids. IEEE Transactions on Sustainable Energy. 2015 April;6(2):635–643.

[6] Chia-Hung Lin, Chao-Shun Chen, Hui-Jen Chuang, Cheng-Yu Ho. Heuristic rule-based phase balancing of distribution systems by considering customer load patterns. IEEE Transactions on Power Systems. 2005 May;20(2):709–716.

[7] Zhu J, Bilbro G, Chow MY. Phase balancing using simulated annealing. IEEE Transactions on Power Systems. 1999 Nov;14(4): 1508–1513.

[8] Lin C, Chen C, Chuang H, Huang M, Huang C. An Expert System for Three-Phase Balancing of Distribution Feeders. IEEE Transactions on Power Systems. 2008 Aug;23(3):1488–1496.

[9] Soltani S, Rashidinejad M, Abdollahi A. Stochastic Multiobjective Distribution Systems Phase Balancing Considering Distributed Energy Resources. IEEE Systems Journal. 2017;PP(99):1–12.

[10] Agrawal A, Verschueren R, Diamond S, Boyd S. A Rewriting System for Convex Optimization Problems. Journal of Control and Decision. 2018;5(1):42–60.

[11] Nesterov Y. Introductory Lectures on Convex Programming Volume I: Basic course. Springer; 2008.

[12] Nesterov Y, Nemirovskii A. Interior point polynomial algorithms in convex programming. vol. 1 of 10. 1st ed. Philadelphia: SIAM studies in applied mathematics; 1994.

[13] Bertsekas D. Convex Optimization Algorithms. Massachusetts Institute of Technology, Athena Scientific, Belmont, Massachusetts; 2015.

[14] Rockafellar T. Lagrange Multipliers and Optimality. SIAM Review. 1993;35(2):183–238.

[15] Slootweg JG, de Haan SWH, Polinder H, Kling WL. General model for representing variable speed wind turbines in power system dynamics simulations. IEEE Transactions on Power Systems. 2003;18(1): 144–151.

[16] Axler S, Gehring F, Ribet K. Linear algebra done wright. 2nd ed. NY: Springer; 2009.

[17] Boyd S, Vandenberhe L. Convex optimization. Cambridge university press; 2004.

[18] Takahashi W. Introduction to Nonlinear and Convex Analysis. vol. 1. 1st ed. Yokohama: Yokohama Publishers; 2009.

[19] Luenberger D. Optimization by vector space methods. Wiley professional paperback series; 1969.

[20] Luenberger D, Ye Y. Linear and Nonlinear Programming. Springer; 2008.

[21] Hubbard JH, Hubbard BB. Vector Calculus, Linear Algebra, And Differential Forms A Unified Approach. Prentice Hall; 1999.

[22] Grant M, Boyd S, Ye Y. Disciplined Convex Programming. In: Liberti L, Maculan N, editors. Global Optimization: From Theory to Implementation; book series Nonconvex Optimization and its Applications. NY: Springer; 2006. p. 155–210.

[23] CVXPY. CVXPY; 2020. Available from: https://www.cvxpy.org/tutorial/advanced/index.html.

[24] Nocedal J, Wrigth SJ. Numerical optimization. Springer; 2006.

[25] Lee J. A first course in combinatorial optimization. 1st ed. Cambridge: Cambridge university press; 2004.

[26] Goemans MX, Williamson DP. Improved Approximation Algorithms for Maximum Cut and Satisfiability Problems Using Semidefinite Programming. J ACM. 1995 Nov;42(6):1115–1145. Available from: https://doi.org/10.1145/227683.227684.

[27] Poljak S, Tuza Z. The expected relative error of the polyhedral approximation of the maxcut problem. Operations Research Letters. 1994;16(4):191 – 198. Available from: http://www.sciencedirect.com/science/article/pii/016763779490068X.

[28] Blekherman G, Parrillo P, Thomas RA. Semidefinite optimization and convex algebraic geometry. SIAM; 2013.

[29] Anjos MF, Lasserre JB. Handbook on semidefinite, conic and polynomial optimization. Springer; 2012.

[30] Wolkowicz H, Romesh S, Lieven V. Handbook of Semidefinite Programming Theory, Algorithms, and Applications. NY: Springer US; 2000.

[31] Cominetti R, Facchinei F, Lasserre J. Modern Optimization Modelling Techniques. Berlin: Springer Basel; 2012.

[32] Anjos MF, Lasserre JB. Handbook on Semidefinite, Conic and Polynomial Optimization. Springer US; 2012.

[33] Correa-Florez CA, Michiorri A, Kariniotakis G. Optimal Participation of Residential Aggregators in Energy and Local Flexibility Markets. IEEE Transactions on Smart Grid. 2020;11(2):1644–1656.

[34] Bertsimas D, Brown DB, Caramanis C. Theory and Applications of Robust Optimization. SIAM Review. 2011;53(3):464–501. Available from: http://www.jstor.org/stable/23070141.

[35] Birge JR, Louveaux F. Introduction to stochastic programming. 2nd ed. NY: Springer; 2011.

[36] Robinson C, Tompsett DH. Power-system engineering problems with reference to the use of digital computers. Proceedings of the IEE - Part B: Radio and Electronic Engineering. 1956;103(1):26–34.

[37] Basu M. Economic environmental dispatch using multi-objective differential evolution. Applied Soft Computing. 2011;11(2):2845 – 2853. The Impact of Soft Computing for the Progress of Artificial Intelligence. Available from: http://www.sciencedirect.com/science/article/pii/S1568494610002917.

[38] Stott B, Jardim J, Alsac O. DC Power Flow Revisited. IEEE Transactions on Power Systems. 2009;24(3):1290–1300.

[39] Harker DC, Jacobs WE, Ferguson RW, Harder EL. Loss Evaluation; Parts I to V. Transactions of the American Institute of Electrical Engineers Part III: Power Apparatus and Systems. 1954;73(1):709–716.

[40] Burnett KN, Halfhill DW, Shepard BR. A New Automatic Dispatching System for Electric Power Systems [includes discussion]. Transactions of the American Institute of Electrical Engineers Part III: Power Apparatus and Systems. 1956;75(3):1049–1056.

[41] Sörensen K. Metaheuristics—the metaphor exposed. International Transactions in Operational Research. 2015;22(1):3–18. Available from: https://onlinelibrary.wiley.com/doi/abs/10.1111/itor.12001.

[42] Barcelo WR, Rastgoufard P. Dynamic economic dispatch using the extended security constrained economic dispatch algorithm. IEEE

Transactions on Power Systems. 1997;12(2): 961–967.

[43] Li Q, Gao DW, Zhang H, Wu Z, Wang F. Consensus-Based Distributed Economic Dispatch Control Method in Power Systems. IEEE Transactions on Smart Grid. 2019;10(1):941–954.

[44] Coffrin C, Knueven B, Holzer J, Vuffray M. The impacts of convex piecewise linear cost formulations on AC optimal power flow. ArXiv. 2020;0(0):42–60.

[45] Morales-España G, Latorre JM, Ramos A. Tight and Compact MILP Formulation for the Thermal Unit Commitment Problem. IEEE Transactions on Power Systems. 2013;28(4):4897–4908.

[46] Castillo A, Laird C, Silva-Monroy CA, Watson J, O'Neill RP. The Unit Commitment Problem With AC Optimal Power Flow Constraints. IEEE Transactions on Power Systems. 2016;31(6):4853–4866.

[47] Anjos M, Conejo A. Unit commitment in electric energy systems. Boston: NOW Foundations and Trends in Electric Energy Systems,; 2017.

[48] González-Castellanos A, Pozo D, Bischi A. In: Distribution System Operation with Energy Storage and Renewable Generation Uncertainty. Cham: Springer International Publishing; 2020. p. 183–218. Available from: https://doi.org/10.1007/978-3-030-36115-0_6.

[49] Padhy NP. Unit commitment-a bibliographical survey. IEEE Transactions on Power Systems. 2004;19(2):1196–1205.

[50] IHA. 2020-hydropower status report, sector trends and insights. International hydropower association; 2020.

[51] WAPA. small-scale hydroelectric power, a brieff assessment. Western area power administration; 1984.

[52] Wood A, Wollenberg B, Sheble G. Power Generation, Operation, and Control. 3rd ed. NY: Wiley; 2013.

[53] Victorov G. Guidelines for the application of small hydraulic turbines. United Nations Industrial Development Organization; 1986.

[54] Fuentes-Loyola, Quintana VH. Medium-term hydrothermal coordination by semidefinite programming. IEEE Transactions on Power Systems. 2003 November;18:1515–1522.

[55] Castano JC, Garces A, Fosso O. Short-Term Hydrothermal Coordination with Solar and Wind Farms Using Second-Order Cone Optimization with Chance-box Constraints. in press. 2020.

[56] Agarwal SK, Nagrath IJ. Optimal scheduling of hydrothermal systems. Proceedings of the Institution of Electrical Engineers. 1972;119(2):169–173.

[57] Diniz AL, Souza TM. Short-Term Hydrothermal Dispatch With River-Level and Routing Constraints. IEEE Transactions on Power Systems. 2014;29(5): 2427–2435.

[58] Pickard WF. The History, Present State, and Future Prospects of Underground Pumped Hydro for Massive Energy Storage. Proceedings of the IEEE. 2012;100(2):473–483.

[59] Luo X, Wang J, Dooner M, Clarke J. Overview of current development in electrical energy storage technologies and the application potential in power system operation. Applied Energy. 2015;137:511 – 536. Available from: http://www.sciencedirect.com/science/article/pii/S0306261914010290.

[60] Suul JA. Variable Speed Pumped Storage Hydropower Plants for Integration of Wind Power in Isolated Power Systems. Renewable Energy, T J Hammons, InTech; 2009.

[61] Redondo NJ, Conejo AJ. Short-term hydro-thermal coordination by Lagrangian relaxation: solution of the dual problem. IEEE Transactions on Power Systems. 1999;14(1):89–95.

[62] Wong KP, Wong YW. Short-term hydrothermal scheduling part. I. Simulated annealing approach. IEE Proceedings - Generation, Transmission and Distribution. 1994;141(5):497–501.

[63] Gil E, Bustos J, Rudnick H. Short-term hydrothermal generation scheduling model using a genetic algorithm. IEEE Transactions on Power Systems. 2003;18(4):1256–1264.

[64] van Ackooij W, Finardi EC, Ramalho GM. An Exact Solution Method for the Hydrothermal Unit Commitment Under Wind Power Uncertainty With Joint Probability Constraints. IEEE Transactions on Power Systems. 2018;33(6):6487–6500.

[65] Bruninx K, Dvorkin Y, Delarue E, Pandžić H, D'haeseleer W, Kirschen DS. Coupling Pumped Hydro Energy Storage With Unit Commitment. IEEE Transactions on Sustainable Energy. 2016;7(2):786–796.

[66] Terry L, Pereira M, Araripe-Neto T, Silva L, Sales P. Coordinating the Energy Generation of the Brazilian National Hydrothermal Electrical Generating System. INFORMS Journal on Applied Analytics. 1986;p. 361–379.

[67] Finardi E, Decker B, de Matos V. An Introductory Tutorial on Stochastic Programming Using a Long-term Hydrothermal Scheduling Problem. Journal of Control Automation and Electric Systems. 2013;24:361–379.

[68] Garces A. Uniqueness of the power flow solutions in low voltage direct current grids. Electric Power Systems Research. 2017;151:149 – 153. Available from: http://www.sciencedirect.com/science/article/pii/S0378779617302298.

[69] Ochoa LF, Wilson DH. Angle constraint active management of distribution networks with wind power. In: 2010 IEEE PES Innovative Smart Grid Technologies Conference Europe (ISGT Europe); 2010. p. 1–5.

[70] Garces A, Ramirez D, Mora J. A Wirtinger Linearization for the Power Flow in Microgrids. Presented in 2019 IEEE Power and Energy Society General Meeting, Atlanta. 2019 Aug;.

[71] Dommel HW, Tinney WF. Optimal Power Flow Solutions. IEEE Transactions on Power Apparatus and Systems. 1968;PAS-87(10):1866–1876.

[72] Peschon J, Bree DW, Hajdu LP. Optimal power-flow solutions for power system planning. Proceedings of the IEEE. 1972;60(1):64–70.

[73] Gómez Expósito A, Romero Ramos E. Reliable load flow technique for radial distribution networks. IEEE Transactions on Power Systems. 1999;14(3):1063–1069.

[74] Torres GL, Quintana VH. An interior-point method for nonlinear optimal power flow using voltage rectangular coordinates. IEEE Transactions on Power Systems. 1998;13(4):1211–1218.

[75] Garces A. A Linear Three-Phase Load Flow for Power Distribution Systems. IEEE Transactions on Power Systems. 2016 Jan;31(1):827–828.

[76] Low SH. Convex Relaxation of Optimal Power Flow—Part I: Formulations and Equivalence. IEEE Transactions on Control of Network Systems. 2014 March;1(1):15–27.

[77] Low SH. Convex Relaxation of Optimal Power Flow—Part II: Exactness. IEEE Transactions on Control of Network Systems. 2014;1(2):177–189.

[78] Molzahn DK, Hiskens IA. A Survey of Relaxations and Approximations of the Power Flow Equations. now; 2019. Available from: https://ieeexplore.ieee.org/document/8635446.

[79] Li J, Liu F, Wang Z, Low SH, Mei S. Optimal Power Flow in Stand-Alone DC Microgrids. IEEE Transactions on Power Systems. 2018;33(5): 5496–5506.

[80] Lavaei J, Low SH. Zero Duality Gap in Optimal Power Flow Problem. IEEE Transactions on Power Systems. 2012;27(1):92–107.

[81] Molzahn DK, Hiskens IA. Convex Relaxations of Optimal Power Flow Problems: An Illustrative Example. IEEE Transactions on Circuits and Systems I: Regular Papers. 2016;63(5):650–660.

[82] Madani R, Sojoudi S, Lavaei J. Convex Relaxation for Optimal Power Flow Problem: Mesh Networks. IEEE Transactions on Power Systems. 2015;30(1):199–211.

[83] Chis M, Salama MMA, Jayaram S. Capacitor placement in distribution systems using heuristic search strategies. IEE Proceedings - Generation, Transmission and Distribution. 1997;144(3):225–230.

[84] Civanlar S, Grainger JJ, Yin H, Lee SSH. Distribution feeder reconfiguration for loss reduction. IEEE Transactions on Power Delivery. 1988;3(3): 1217–1223.

[85] Lavorato M, Franco JF, Rider MJ, Romero R. Imposing Radiality Constraints in Distribution System Optimization Problems. IEEE Transactions on Power Systems. 2012;27(1):172–180.

[86] Gil-González W, Garces A, Montoya OD, Hernández JC. A Mixed-Integer Convex Model for the Optimal Placement and Sizing of Distributed Generators in Power Distribution Networks. Applied Sciences. 2021;11(2). Available from: https://www.mdpi.com/2076-3417/11/2/627.

[87] Divan D, Kandula P. Increasing solar hosting capacity is the key to sustainability. In: 2016 First International Conference on Sustainable Green Buildings and Communities (SGBC); 2016. p. 1–5.

[88] Teodorescu R, Liserre M, Rodriguez P. Grid Converters for Photovoltaic and Wind Power Systems. 1st ed. NY: IEEE Power Engineering Society, NJ, Wiley-Interscience; 2011.

[89] Garces A. A Linear Three-Phase Load Flow for Power Distribution Systems. Power Systems, IEEE Transactions on. 2015;PP(99):1–2.

[90] Gallego RA, Monticelli AJ, Romero R. Optimal capacitor placement in radial distribution networks. IEEE Transactions on Power Systems. 2001;16(4):630–637.

[91] Kefayat M, Ara AL, Niaki SN. A hybrid of ant colony optimization and artificial bee colony algorithm for probabilistic optimal placement and sizing of distributed energy resources. Energy Conversion and Management. 2015;92:149–161.

[92] Su X, Masoum MAS, Wolfs PJ. PSO and Improved BSFS Based Sequential Comprehensive Placement and Real-Time Multi-Objective Control of Delta-Connected Switched Capacitors in Unbalanced Radial MV Distribution Networks. IEEE Transactions on Power Systems. 2016;31(1):612–622.

[93] Sorensen K. Metaheuristics—the metaphor exposed. International Transactions in Operational Research. 2015;22(1):3–18. Available from: https://onlinelibrary.wiley.com/doi/abs/10.1111/itor.12001.

[94] Dubey A, Santoso S. On Estimation and Sensitivity Analysis of Distribution Circuit's Photovoltaic Hosting Capacity. IEEE Transactions on Power Systems. 2017 July;32(4):2779–2789.

[95] Torquato R, Salles D, Oriente Pereira C, Meira PCM, Freitas W. A Comprehensive Assessment of PV Hosting Capacity on Low-Voltage Distribution Systems. IEEE Transactions on Power Delivery. 2018 April;33(2):1002–1012.

[96] Al-Saadi H, Zivanovic R, Al-Sarawi SF. Probabilistic Hosting Capacity for Active Distribution Networks. IEEE Transactions on Industrial Informatics. 2017 Oct;13(5):2519–2532.

[97] Gensollen N, Horowitz K, Palmintier B, Ding F, Mather B. Beyond Hosting Capacity: Using Shortest-Path Methods to Minimize Upgrade Cost Pathways. IEEE Journal of Photovoltaics. 2019 July;9(4):1051–1056.

[98] Garces A, Molinas M, Rodriguez P. A generalized compensation theory for active filters based on mathematical optimization in ABC frame. Electric Power Systems Research. 2012;90:1–10. Available from: https://www.sciencedirect.com/science/article/pii/S0378779612000806.

[99] Akagi H, Watanabe EH, Aredes M. Instantaneous Power Theory and Applications to Power Conditioning. 1st ed. NY: IEEE Power Engineering Society, NJ, Wiley-Interscience; 2007.

[100] Czarnecki LS. Instantaneous reactive power p-q theory and power properties of three-phase systems. IEEE Transactions on Power Delivery. 2006;21(1):362–367.

[101] Herrera RS, Salmeron P. Instantaneous Reactive Power Theory: A Comparative Evaluation of Different Formulations. IEEE Transactions on Power Delivery. 2007;22(1):595–604.

[102] Baran ME, Wu FF. Network reconfiguration in distribution systems for loss reduction and load balancing. IEEE Transactions on Power Delivery. 1989;4(2):1401–1407.

[103] Monticelli A. Electric power system state estimation. Proceedings of the IEEE. 2000;88(2):262–282.

[104] Ardakanian O, Wong VWS, Dobbe R, Low SH, von Meier A, Tomlin CJ, et al. On Identification of Distribution Grids. IEEE Transactions on Control of Network Systems. 2019;6(3):950–960.

[105] Farajollahi M, Shahsavari A, Mohsenian-Rad H. Topology Identification in Distribution Systems Using Line Current Sensors: An MILP Approach. IEEE Transactions on Smart Grid. 2020;11(2):1159–1170.

[106] Schweppe FC, Wildes J. Power System Static-State Estimation, Part I: Exact Model. IEEE Transactions on Power Apparatus and Systems. 1970;PAS-89(1):120–125.

[107] Zhao J, Gómez-Expósito A, Netto M, Mili L, Abur A, Terzija V, et al. Power System Dynamic State Estimation: Motivations, Definitions, Methodologies, and Future Work. IEEE Transactions on Power Systems. 2019;34(4):3188–3198.

[108] Zhu H, Giannakis GB. Power System Nonlinear State Estimation Using Distributed Semidefinite Programming. IEEE Journal of Selected Topics in Signal Processing. 2014;8(6):1039–1050.

[109] Madani R, Lavaei J, Baldick R. Convexification of Power Flow Equations in the Presence of Noisy Measurements. IEEE Transactions on Automatic Control. 2019;64(8):3101–3116.

[110] Wu FF, Monticelli A. Network Observability: Theory. IEEE Transactions on Power Apparatus and Systems. 1985;PAS-104(5):1042–1048.

[111] Bai H, Zhang P, Ajjarapu V. A Novel Parameter Identification Approach via Hybrid Learning for Aggregate Load Modeling. IEEE Transactions on Power Systems. 2009;24(3):1145–1154.

[112] Garces A, Gil-González W, Montoya OD, Chamorro HR, Alvarado-Barrios L. A Mixed-Integer Quadratic Formulation of the Phase-Balancing Problem in Residential Microgrids. Applied Sciences. 2021;11(5). Available from: https://www.mdpi.com/2076-3417/11/5/1972.

[113] Bazaraa M, Jarvis J, Sherali H. Linear programming and network flows. Wiley; 2010.

[114] Jinxiang Zhu, Mo-Yuen Chow, Fan Zhang. Phase balancing using mixed-integer programming [distribution feeders]. IEEE Transactions on Power Systems. 1998;13(4):1487–1492.

[115] Gellings CW. The concept of demand-side management for electric utilities. Proceedings of the IEEE. 1985;73(10):1468–1470.

[116] Meyabadi AF, Deihimi MH. A review of demand-side management: Reconsidering theoretical framework. Renewable and Sustainable Energy Reviews. 2017;80:367 – 379. Available from: http://www.sciencedirect.com/science/article/pii/S1364032117308481.

[117] Deng R, Yang Z, Chow M, Chen J. A Survey on Demand Response in Smart Grids: Mathematical Models and Approaches. IEEE Transactions on Industrial Informatics. 2015;11(3):570–582.

[118] Hu RL, Skorupski R, Entriken R, Ye Y. A Mathematical Programming Formulation for Optimal Load Shifting of Electricity Demand for the Smart Grid. IEEE Transactions on Big Data. 2020;6(4):638–651.

[119] Zhu J, Bilbro G, Mo-Yuen Chow. Phase balancing using simulated annealing. IEEE Transactions on Power Systems. 1999;14(4):1508–1513.

[120] Castaño JC, Garcés A, Rios MA. In: Phase Balancing in Power Distribution Grids: A Genetic Algorithm with a Group-Based Codification. Cham: Springer International Publishing; 2020. p. 325–342. Available from: https://doi.org/10.1007/978-3-030-36115-0_ 11.

[121] Li K, Bian H, Liu C, Zhang D, Yang Y. Comparison of geothermal with solar and wind power generation systems. Renewable and Sustainable Energy Reviews. 2015;42:1464–1474. Available from: https://www.sciencedirect.com/science/article/pii/S1364032114008740.

Index

Mathematical Programming for Power Systems Operation: From Theory to Applications in Python. First Edition. Alejandro Garcés.
© 2022 by The Institute of Electrical and Electronics Engineers, Inc. Published 2022 by John Wiley & Sons, Inc.

Printed and bound by CPI Group (UK) Ltd, Croydon, CR0 4YY
05/11/2021
03090625-0002

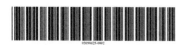